Theodore Wood, John George Wood

Hughes's Anecdotal Natural History

Theodore Wood, John George Wood

Hughes's Anecdotal Natural History

ISBN/EAN: 9783337149604

Printed in Europe, USA, Canada, Australia, Japan

Cover: Foto ©berggeist007 / pixelio.de

More available books at **www.hansebooks.com**

Hughes's
Anecdotal Natural History

BY REV. J. G. WOOD
AND
THEODORE WOOD

WM. C. G. LUDFORD,
Hill,
SUTTON COLDFIELD.

William Sudford

Christmas

1882.

HUGHES'S

ILLUSTRATED

Anecdotal Natural History.

BY

THE REV. J. G. WOOD, M.A., F.L.S.,

Author of "Homes Without Hands," &c., &c.

AND

THEODORE WOOD, M.E.S.,

Joint author of "The Field Naturalist's Note-Book."

London:

JOSEPH HUGHES, PILGRIM STREET,

LUDGATE HILL, E.C.

1882.

CONTENTS.

ANECDOTAL NATURAL HISTORY.

—o—

No. I.—THE SQUIRREL TRIBE.

ALMOST everybody is familiar with the common squirrel (*Sciurus Europæus*), that reddish-brown animal with the bushy tail which is so plentiful in nearly all our woods and forests, where it sometimes works considerable mischief. And even those people who have never been fortunate enough to see it in its native haunts, springing from tree to tree, and gambolling merrily among the branches, must have noticed the unfortunate specimens exhibited for sale by the itinerant hawkers who pervade the streets of the metropolis and other large towns.

Besides the British species there are many other squirrels found in different parts of the world, Australia being the only continent where none are known to exist.

Particular attention is drawn to this point, because in nearly all travellers' accounts of Australia we read of the flying squirrel among the quadrupeds inhabiting that land. In fact, however, the so-called squirrels are not squirrels, nor even rodents, but are Marsupials, belonging to the great group of Phalangists.

The mistake arises from the natural errors made by travellers and colonists, who name every creature they see after the inhabitants of their own country.

Our Australian colonists talk and write with marvellous composure of wolves, bears, monkeys, bats, cats, squirrels, rats, and mice as inhabiting the country. Similarly, American writers sadly bewilder the tyro in zoology by mentioning American buffaloes, lions, tigers, and panthers, all these creatures being strictly confined to the old world. The 'Robin' again, so often mentioned in American literature, is not our English redbreast, as is usually assumed, but an erroneous name for the migratory thrush, a bird far larger than the redbreast and belonging to a totally different group.

The squirrels belong to the great division of the rodents, which comprise nearly a third of all the known mammalia. The animals of this group are distinguished by the possession of two powerful chisel-edged incisor teeth in each jaw, formed for cutting or gnawing away hard substances, and which are replaced by fresh material as fast as they are worn away. The power of these incisor teeth is strikingly exemplified in the beaver, which has been known to gnaw its way through logs no less than eighteen inches in diameter.

Were there not some means of replacing these teeth as quickly as they are worn down, the death of the animals would speedily follow, as they would shortly starve from their inability to procure food. In order to avoid this, the teeth are continually forced forward from the jaw by the formation of fresh substance at the base, which is secreted by a pulpy substance at the root of the tooth. Now, as this growth takes place whether the teeth are used or not, it follows that, unless they were in constant use, they would soon increase to an inordinate length, and before very long would project from the mouth. Such an event does occasionally occur, when by some accident one of the incisor teeth has been broken short off. The corresponding tooth in the other jaw, finding no resistance to its growth, continues to increase until it sometimes forms a perfect circle outside the mouth, usually resulting in the death of its owner by preventing it from feeding.

Without some means, however, of preserving the chisel-like sharpness of these teeth, the mere replacement of wasted substance would be of little use. In order to obtain the desired result, the teeth are constructed after a very singular fashion.

The body of the tooth is composed of pure ivory, coated on the outer surface with a thin layer of enamel, which being of a very much harder nature, is not worn away with the same rapidity. Besides this, the ivory nearest the enamel is harder than the rest, and the softer parts being easiest worn down, the edge of the tooth always keeps the same proportions, the actual cutting being performed with the edge of the enamel.

Our carpenter's chisels are constructed on exactly the same principle, the chief portion of them being composed of soft iron, while a very thin plate of steel is laid along the back and forms the cutting edge of the tool.

So much for the characteristics of the rodents as a whole. Now for those of that group which are known as squirrels.

The true squirrels are scientifically known as *Sciuridæ*, or shadow-tails, a title derived from two Greek words, the former signifying a shade and the latter a tail. This refers to the habit in these animals of carrying the tail over the back, as though to protect the body from the rays of the sun, a position always adopted except when running or leaping. They are remarkable among the rodents as possessing particu-

larly perfect clavicles or collar-bones, which enable
them to use the fore-paws to a certain extent after the
manner of hands. This is especially the case in carrying food to the mouth, when the paws are managed
with extraordinary dexterity. In eating a nut, for example, a squirrel, by the aid of its fore-paws and teeth
alone, will break the shell and peel the kernel to the
full as successfully as a skilful human being furnished
with crackers and penknife. And it does so as
follows. Holding the nut close to the teeth, it gnaws
away at the point of the fruit until it fairly pierces the
outer shell. It then dexterously inserts the edge of
the upper teeth into the aperture, and splits away the
shell just as an idle boy does with his knife when
opening nuts in school hours. By means of its hand-
like paws, it then holds the kernel against its upper
teeth, and rapidly turning it round and round, strips
off the whole of the peel before beginning to devour
it.

With the exception of the jaws and the adjacent
parts of the head, the skeleton is exceedingly light
and slender in order to suit it to the rapid movements
of the animal.

To furnish the squirrels with the means of ascending
the trees in which they spend the greater portion of
their existence, the long toes are provided with sharp,
curved claws, which can be inserted into the smallest
crevices of the bark, and thus secure a firm foothold.

The rapidity and ease of their aerial motions is
something astounding. A squirrel will gallop up a
perpendicular tree trunk fully as fast as a cat can run
on level ground, and will throw itself from branch to
branch with the most perfect recklessness. And even
if it misses its mark, it simply extends its limbs and
allows itself to fall to the ground from most alarming
heights, never seeming to be in the least damaged by its

Squirrel's Summer Nest.

rapid descent. One would imagine that the creature must infallibly be dashed to pieces, but, almost before one can look round, the squirrel is off to the nearest tree trunk, where it is quickly lost to sight amongst the foliage.

As a refuge both by day and night, and also as a place wherein to rear their young, the squirrels always construct a large and comfortable nest, generally in the fork of some lofty branch, and always carefully concealed from sight. The materials are chiefly leaves, grass, and moss, woven together in a most beautiful manner, and forming a perfect protection against the rain. The old nests may sometimes be seen in winter when they are no longer surrounded by foliage, and stand out boldly among the naked boughs.

In these nests the young, three or four at a litter, are brought up, and remain until they are old enough to shift for themselves.

At the approach of winter the squirrels, not being able to find their food during the cold months, lay up stores for use during that period. Their provisions consist generally of nuts of various sorts, grain, beech-mast, and so on, and are hidden away as a rule in holes at the roots of trees in the neighbourhood of their winter habitation. It is a somewhat curious fact that the squirrel never includes a bad nut in these hoards.

This winter retreat is also a nest of much the same character as the former, but hidden away in holes in trees, the junctions of large branches with the trunk, and similar localities. As soon as the cold weather fairly arrives, the squirrels, having previously laid up their winter stores, repair to this hiding-place, and there pass the time until spring, when they again return to their summer haunts. For the greater portion of the interven-

Squirrel's Winter Nest

ing period they lie in a state of complete unconsciousness, something more than sleep and less than
death, and which can be only partially described by
the word 'trance.' This state is not, as is generally
imagined, produced by intense cold; for should the
temperature fall below 32° Fahrenheit, or 'freezing-
point,' a hibernating animal is first awakened and
then killed by the frost. Those animals, therefore,
which pass the winter in a state of unconsciousness,
always select a retreat where they are sheltered from
the direct action of the elements, and where the
temperature is only slightly varied.

While in the hibernating condition respiration entirely ceases, and the animal could be kept for hours
under water, or in a vessel of carbonic acid gas,
without the slightest effect. The circulation of the
blood is greatly retarded, and digestion, at any rate
in those animals which spend the whole winter in
unconsciousness, is wholly at a standstill.

The squirrel, however, is only a partial hibernator,
and is obliged to leave its hiding-place three or four
times during the winter months in search of food. For
this purpose it always selects a milder day than usual,
and on such occasions may be seen repairing to its
stores for a meal before again proceeding to its place
of refuge. These hoards seem always to be more
than sufficient for its wants, and the remainder of the
provender is left to its fate. Sometimes the nuts, etc.,
take root, and the squirrel is often the cause of trees
springing up in unexpected places.

In the far north the cold of the winter has a very
marked effect upon the fur of these animals, which
from a rich brownish red becomes of a pale grey hue.
And even in our own climate there is a perceptible
change in the colour of the fur at different periods of
the year.

It is not generally known that squirrels can use their paws as paddles, and by their aid swim to considerable distances.

In some parts of the eastern coasts of Scotland, where innumerable armlets of the sea run many miles inland, the squirrels are in the habit of swimming across from shore to shore, and making rapid progress through the water. A friend was an eye-witness of this habit. Being in a boat, he came upon the little creature as it was swimming, and took it into the boat.

As it was tired with its journey, it allowed itself to be caught without any difficulty. But as soon as it felt itself rested, it sprang back into the water and swam steadily to land

Leaving this country for America, we find there are many kinds of squirrels found in that country, several of which are very interesting. For example, there is the Grey Squirrel (*Sciurus cinereus*) of North America, which is extremely abundant, and often works very great damage to the growing crops. In Pennsylvania, more than a century ago, these animals were found to be so very destructive that a Government reward of threepence per head was offered for all that were killed. In a single year, a sum of no less than £8000 was expended in redeeming this promise, which gives a grand total of 640,000 squirrels destroyed.

Now, although these animals wrought such terrible mischief to the agriculturist, and by their numbers and perseverance ruined the result of his labours, in their native forests they were in their proper situation, doing the work for which they were sent into the world. But as soon as man arrived upon the scene, cleared away the jungle, and laid out the ground for cultivation, he upset the balance of nature : the squirrels were no longer required, and became a plague instead of a benefit, making it necessary to thin their number. Even our own squirrel is occasionally the cause of much damage, especially in young plantations, where it has a habit of nibbling off the topmost shoots, and so stopping the growth of the tree. Nor does it confine its depredations to vegetable life, for it is by no means uncommon to find a bird's nest ransacked by the animal, and the eggs or young ones devoured, as the case may be. It is probable that the squirrel is the real delinquent in many a case of nest-robbing when the blame falls on the shoulders of thoughtless schoolboys.

It occasionally happens with the grey squirrels, that, having pretty thoroughly devastated a neighbourhood, and finding winter approaching, they are unable to lay up a sufficient stock of food on which to subsist until the spring. Knowing instinctively that if they remain in their present locality they must inevitably die of starvation, they migrate in vast numbers, after the fashion of the lemming in Northern Europe, allowing nothing to check their course, climbing over instead of avoiding any obstacle, such as a wall or house, and leaving nothing eatable behind them. Every blade of grass and every green thing disappears, and the transit of one of these hosts leaves the country in much the same condition as if a swarm of locusts had passed over it.

Then there is the Black Squirrel (*Sciurus niger*), which, though not nearly so numerous as the preceding species, is still far from uncommon. It is a curious circumstance that the black and the grey squirrels seem unable to live in company, and as soon as the latter animal shows itself, the former disappears. As its name implies, this animal is of a uniform black hue, and from the fineness of the fur, is much sought after for the sake of the skin.

There is a somewhat strange-looking squirrel inhabiting Borneo, and which is popularly known as the Long-eared Squirrel (*Sciurus macrotis*). It is thus named on account of the singularly long fringe of hair with which the ears are decorated, which is of a dark brown colour, and pretty well two inches in length. The tail also is remarkably bushy.

Besides all these, there are some animals, none of them inhabiting this country, however, which are generally known as 'flying' squirrels on account of a singular modification of structure. This is found in the skin of the flanks, which is developed to a large extent, almost hiding the paws in its folds when the creature is at rest. When in motion, however, and particularly during the tremendous leaps which these animals make from branch to branch, the legs are stretched out as far as possible, the loose skin acts as a parachute, and the squirrel is enabled to pass through a much greater distance than would otherwise have been possible.

The petaurists of Australia possess a similar development of the skin, and use it for a similar purpose.

One of the best known of the flying squirrels is the Taguan (*Pteromys petaurista*) of India, which is of a brownish colour, varying from deep chestnut along the back to a greyish white on the under surface of the body. The tail is long and bushy, and very much darker in colour. The whole length of the animal is nearly three feet.

Leaving the true squirrels, we come to a very closely allied group of rodents, the members of which, however, construct their habitations beneath the earth instead of among the branches of trees, and but rarely leave the ground. These are called ground squirrels, and are furnished with cheek-pouches, which the true squirrels do not possess. The object of these we shall presently see.

The Hackee, Chipping Squirrel, or Chipmuck, as it is indifferently termed (*Tamias Lysteri*), of North America is one of the most widely known of these quadrupeds. It is a very pretty little creature, rather less than a foot in total length, of a brownish-grey hue, with five black and two yellow stripes running longitudinally along the back. The under surface is of a fine white.

The hackee is very abundant over a great portion of North America, and may be seen almost everywhere dashing in and out of the underwood with a rapidity which has caused its movements to be compared to those of the wren.

By way of habitation, and also as a refuge from its numerous foes, it constructs rather complicated burrows below the surface of the earth. For this purpose it generally selects some protected situation, such as the roots of a large tree, the side of a bank, or the foot of a hedge. A winding tunnel, usually of considerable length, leads to the dwelling chamber, or nest, where the young are brought up, and from this run galleries leading to other chambers which are used for storehouses. In these a most astonishing quantity of food is laid up. In a single nest, we are told, were found no less than a peck of acorns, two quarts of buckwheat, a quart of beaked nuts, some grass seeds, and a quantity of maize, with which the interstices were filled up.

It seems almost incredible that so small an animal as the hackee can lay up so large a quantity of food, but such is the case, and it is in carrying the provisions to the burrow that the use of the cheek pouches is found. It can, of course, carry only a small quantity at a time. Four beaked nuts, for example, constitute the load for a single journey, three being packed into the pouches, and the fourth carried between the teeth. In order to prevent the nut from hurting its mouth, the hackee invariably bites off the sharp beak of the fruit before consigning it to the pouches.

When thus loaded, the animal presents a very curious appearance, bearing a ludicrous resemblance to a human being suffering from a very bad attack of mumps.

The well-supplied condition of the hackee's larder is widely known, and in times of scarcity the natives repair to the burrows, and, digging out the contents, find the materials for a hearty meal.

When pursued by one of its numerous enemies, the chipping squirrel always takes refuge in its burrow, trusting in the inability of its pursuer to follow it down the complicated windings of its narrow tunnel. There is one foe, however, to whom this is no obstacle, and which follows the hackee to the very end of the burrow, there making a meal of the unfortunate owner. This enemy is found in the stoat, which sometimes kills the whole of the occupants of a burrow merely for the gratification of sucking their blood.

All three of the popular names are applied to this animal on account of the curious cry, which somewhat resembles the clucking of newly-hatched chickens. The scientific title, *Tamias*, is a Greek word, signifying a storekeeper, the application of which is at once apparent.

B

No. II.—THE CAMEL.

THE place of the horse is supplied, and more than supplied, in the East by the Camel, the 'Ship of the Desert,' as the Arabs poetically term it, whose structure is singularly adapted to the nature of the country which it has to traverse. A horse could not pass through the sandy deserts, beneath the fierce heat of a tropical sun, bearing the heavy load of the camel, and dispensing almost entirely with food and water, especially the latter, for days together; and would, in fact, find it by no means easy, even if entirely unencumbered, to make its way through the treacherous soil, in which it would sink for several inches at every step. But the camel has evidently been constructed with an especial view to the exigencies of the desert traffic, and can therefore perform work which would be quite impossible to any other animal.

There are two species of camel, the chief of which is the true camel of Arabia, bearing one hump only upon its back. This is the animal most in use as a beast of burden. The second species, the Bactrian camel, or Mecheri, which possesses a second hump, is almost entirely used for riding purposes.

Now, when we consider the nature of the soil over which the Arabian camel has to pass, it is evident that there must be some provision in the foot of the creature in order to enable it to find a firm foothold. So, instead of the single hard hoof which is found in the horse, the foot of the camel is provided with two broad toes, furnished with large, soft, and very elastic cushions beneath, so as to afford a large surface, and prevent the animal from sinking in the loose sand.

Then, as the animal must be loaded and unloaded when in a kneeling position, the knees and breast are furnished with thick callous pads, so that the skin is not injured by the contact with the rough ground.

A curious part of the structure of the camel is found in the hump, which is not, as many think, a malformation of the back, for the spine is as straight as that of any other animal, but is merely a fleshy and fatty protuberance, connected in some strange way with the health of its owner. The Arab always judges the state of his camel by the hump, and will not allow the animal to start upon a long journey unless the hump is in perfect condition.

The chief use of the hump seems to lie in its power of nourishing the animal when other food is scarce. During a long desert journey, for example, the camels are very sparingly fed, and appear to subsist upon the nourishment derived from the hump, which gradually diminishes in size, until at the end of the journey it is scarcely visible, not regaining its former proportions until after two or three months of careful feeding.

In a somewhat similar manner, the hibernating animals exist without food for several months. They accumulate a quantity of fat largely in excess of the normal amount, and by the absorbing of the superabundant fat into the body, are enabled to supply the waste of tissue.

Another noticeable point in the structure of the camel is to be found in the formation of the eyes and nostrils, the former of which are provided with long lashes, while the latter can be closed at will, thus preventing the admission of grains of sand during the storms so common in the desert.

Halt in the Desert

Perhaps the most extraordinary part of the formation of the camel lies in its power of storing up sufficient water within the stomach to last it for several days without again drinking. To understand this properly, we must enter somewhat more carefully into the details of the internal anatomy.

Like that of the sheep and other ruminating animals, the stomach of the camel is divided into no less than four portions, which, though each is a separate cavity, are connected with each other. Into the first and largest of these, usually known as the paunch, or *rumen*, the food passes as soon as swallowed, and before it is masticated. There it remains until a convenient opportunity arises for chewing it, when it is returned to the mouth and thoroughly macerated. Before this process takes place, however, it passes into the second stomach or *reticulum*, which consists of a number of polygonal cells, in which the food is formed into a number of smooth balls. This is the 'honeycomb tripe' of butchers. Thence it is expelled into the œsophagus, or gullet, which opens both into the first and second stomachs, and is carried by the contraction of the spiral muscles composing that tube into the mouth, where it is masticated at leisure.

As soon as it is thoroughly chewed, the food is once more swallowed, and this time passes directly through the first and second stomachs into the third, or *psalterium*, the walls of which are composed of very numerous folds, not unlike the leaves of an uncut book. This is the 'manyplies' or 'manyplus' tripe of butchers, presenting a very large surface to the food which is here prepared for admission to the fourth stomach.

This, which is scientifically known by the name of *abomasus*, and by butchers called the 'red,' is the true digestive stomach, the other three being only employed in the prior preparation of the food. In this division of the stomach the gastric juice is

secreted.   In the calf it is called 'rennet,' and is used
for curdling milk in the preparation of cheese.

While the animals of this class, the Ruminants, are
still young, and are fed by their mother's milk, the fourth
stomach only is fully developed, the other three not

Arabian Camel.

being required until the animal is old enough to find
its nourishment in vegetable substances.

This structure is common to all the ruminating
animals, but in the camel there is found to be a still
further development.   The polygonal cells which are

found in the reticulum, and in part of the paunch, are very large in proportion, and are surrounded by muscular bands, enabling them to be closed at will, their contents not mingling with the food contained in the stomach. In these cells is reserved the water drunk by the animal, a small quantity of which can be released and allowed to flow into the stomach as occasion requires, the rest remaining in the reservoirs until needed. By this arrangement a camel is able to store up sufficient liquid for five or six days, a provision of the greatest service in crossing the hot arid desert where water is unprocurable.

It would seem that by practice the camel is enabled to store away a larger quantity of water than had previously been the case, for an old and experienced animal will contrive to lay up half as large again a stock as it could when young and unused to desert travelling.

It has sometimes happened in a caravan that the water has run short, and the only alternative has been to kill some of the camels in order to obtain the water contained in their stomachs. It is then found to be of a pale greenish colour, and very unpleasant to the taste. Yet it is preferable to dying of thirst, and is really hardly more disagreeable than the water contained in the leathern bags carried on the camel's backs, which is heated by the sun, besides tasting very strongly of the tar with which the seams are dressed.

This structure presents a singular analogy to the blood-reservoir of the whale, by which it is enabled to spend a considerable time beneath the surface of the water. A large supply of blood being purified and aerated in the lungs, is stored away in a mass of blood-vessels set apart for that purpose, whence a portion is introduced into the circulatory apparatus as it is from time to time required. But for this provision the whale could never spend more than two or three minutes together beneath the surface of the water, the air or aerated blood being quite as necessary to its existence as the water is to that of the camel.

Of the two species of camel, the Arabian is by far the more valuable, being both stronger and more enduring of privation and fatigue than its Bactrian relative. The load of the Arabian camel is usually from five to six hundred pounds in weight, this being the average amount that the creature can carry with ease. It is by no means a swift creature, its pace seldom exceeding two miles and a half in the hour, and often not coming up to even that standard. There is a swifter breed, usually known as the Dromedary, which is chiefly kept for the saddle, and which can keep up a pace of eight or nine miles per hour for twenty hours at a stretch, being to the camel what the racer is to the cart-horse. The motion of the camel is most unpleasant to any one riding it, for it moves both legs of each side together, progressing at a long swinging trot, and jolting its unfortunate rider in the most unmerciful manner. Novices in camel-riding almost always suffer from sickness as badly as if they were in a Channel steamer on a rough day. With some of the faster dromedaries it is even necessary for the rider to swathe his body, from the hips to the arms, in bandages drawn as tightly as possible, before he commences his journey.

There is a mistaken idea that the camel is a very patient, gentle, and docile animal, and that he is very easily managed. In reality, the case is just the reverse, for a more quarrelsome, unruly, and revengeful animal hardly ever existed. No sooner is he unloaded than he begins to fight any of his fellows who may be in the neighbourhood; the loading and unloading are never

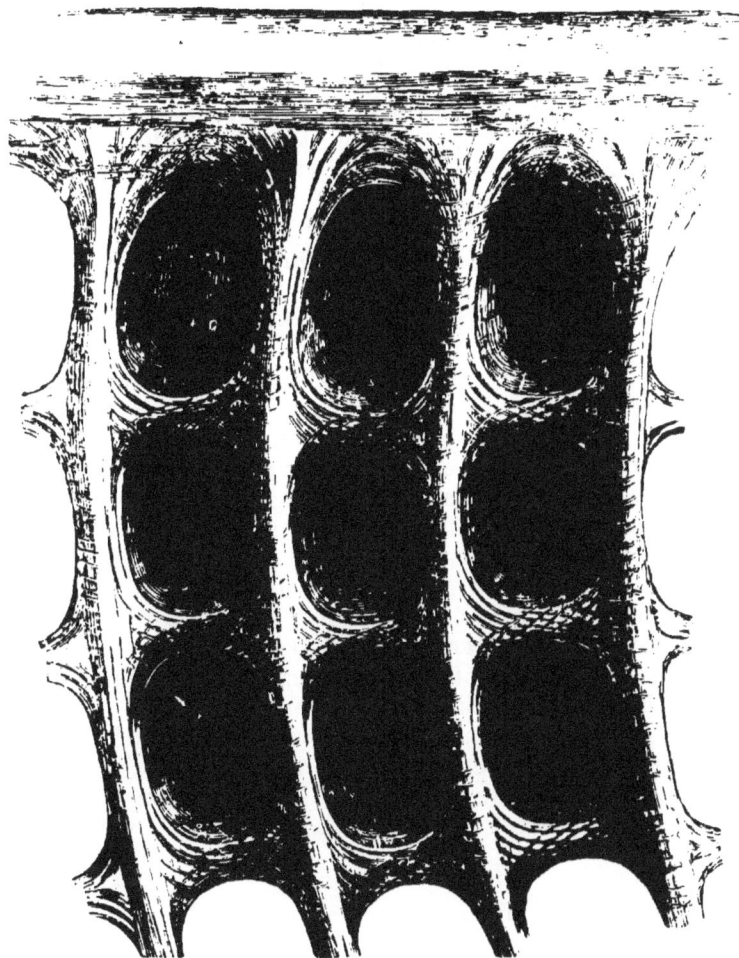

Water Cells in Camel's Stomach

performed without much trouble, and many savage grunts on the part of the camel, and when fairly loaded, his first endeavour is always to get free of his burden, or, failing that, to ruin every article included in it.

Then, at night, when it is time to unload, the cross-grained animal has to be compelled to kneel by main force, and his head tied to his fore legs in such a manner that he cannot rise until the proper time. An experienced traveller says that he has never yet seen a camel in anything but a bad temper, at any rate, to judge by appearances.

Besides its use as a beast of burden, the camel is of service to its masters in various other ways. Its milk, for example, is a standard article of food, and is mostly kept until quite sour, the Arabs considering it to be then a much greater dainty than when it is sweet and fresh. A very inferior kind of butter is churned from the cream by pouring it backwards and forwards in a goat skin for a certain time.

The flesh of the camel is considered a great dainty, but is very seldom eaten, owing to the value of the animal. Occasionally, however, a rich Arab will kill one of his camels, and invite all his friends to a feast, at which the flesh of the slaughtered animal appears as the crowning delicacy.

At certain times of the year the camel sheds its hair, which is collected, and being spun into thread, is used in making garments. Certain portions of it are also utilized for the 'camel's hair pencils' used by artists.

The two-humped or Bactrian camel, which is found throughout Central Asia and China, though now almost entirely used for the saddle, was at one time put to a very curious use. The East India Company formed a regiment of these animals, each being provided with a couple of swivel guns, placed between the humps, and managed by the rider. This company was known as the camel artillery, and was sometimes of considerable service.

The colour of the Bactrian camel is darker than

Bactrian Camel.

that of its Arabian relative, varying from dark brown to a sooty black in hue.

THOUGH the true camels are exclusively confined to the Old World, a closely allied genus, comprising four species, is found in America. These animals are popularly known by the name of Lamas, and differ in many respects from the true camels.

In the first place, the toes of the foot, instead of being connected, as in the camel, are separated, and can be extended at will. This is on account of the rocky and mountainous nature of the localities which they inhabit, and in which the power of moving the toes is necessary in order to give them a firm foothold. All the Llamas are of very much less size than the camel, the Guanaco, the largest, not standing much more than three feet six inches at the shoulder. The hair is long and woolly, and the general aspect of the animals is remarkably like that of an overgrown sheep.

The four species are the Vicugna, the Guanaco, the Yamma, and the Alpaca. The first of these is found in the most mountainous parts of Northern Chili and Batavia, and is valuable on account of its skin, which causes it to be much sought after. In other ways it is entirely useless, as from its wild and untameable nature it cannot be employed as a beast of burden. In colour it is brown, approaching to grey beneath ; the height is about two feet six inches at the shoulder.

The Guanaco, which is found in the more northern regions of Patagonia, is of a reddish-brown colour, the ears and hind legs being grey, and stands about three feet and a half at the shoulder. It lives in herds, varying from ten to forty or more in number, and like the sheep, under the guidance of a single leader, whose orders are always implicitly obeyed. Should this leader be killed or trapped, the flock seem perfectly bewildered, and wander vaguely from place to place, laying themselves open to easy capture by the hunters. The sense of curiosity is very strongly developed in the Guanaco, which, though naturally a wary and timid animal, can be brought within a short distance of a hunter if he lies on his back on the ground and kicks his legs in the air. It is able to swim well, and has often been known to take voluntarily to the water and swim from one island to another.

The Yamma, or Llama, which was formerly used as a beast of burden by the Spaniards in America, is of a variegated brown colour, with long and slender legs. It is now little used in any way, the sheep having replaced it with regard to the wool supply, and its flesh being dark and coarse and seldom eaten.

The fourth species, the Alpaca, or Paco, as it is sometimes termed, is, together with the last-mentioned animal, sometimes thought to be only a domesticated variety of the Guanaco. It is a valuable creature on account of its wool, which is long and silky. A herd of Llamas was even imported into Australia, where it flourished fairly well, yielding a large supply of the valuable wool.

No. III.—BATS.

IN almost every temperate part of the world, but more especially abounding in tropical climates, are found the curious creatures which are popularly known as bats, and scientifically as *cheiroptera*, an appropriate word signifying ' hand-winged ' animals. Australia, however, must be excepted, as the whole of the Australian mammalia belong to the Marsupials.

It is only of late years that their proper position in the scale of creation has been discovered. Before that time, some of the wildest conjectures were made on the subject. As the creatures possessed the power of flight, some authors placed them among the birds, entirely overlooking the differences in structure, which should at once have pointed out their place among the mammals. Some, considering them to be quadrupeds, because they were able to walk upon the ground, though after a rather clumsy fashion, imagined that they must form a connecting link between the mammals and the birds ; and it was not until later discoverers carefully investigated their anatomy that the real position of the bats was arrived at, namely, just after the monkey tribe, and before the cats.

The appearance of the bat is familiar to almost all, the strange membranous wings, enabling their owner to pursue their aerial evolutions with an ease and rapidity not exceeded by any bird, being the first points which arrest the attention.

Though possessing an almost equal power of flight with the birds, the wings of the bat are by no means constructed upon the same principle. Instead of feathers, the wing is composed merely of a membrane tightly stretched between the bones of the fingers, and extending along the sides as far as the tail. In order to fully understand this structure, we must examine the modifications of the skeleton which render it possible.

In the first place, the framework of the wing of the bat is formed merely by the bones of the arm and hand, which, more especially those farthest from the body, are elongated to a wonderful extent, the middle finger being actually of greater length than the whole head and body of the animal. The only exception is the thumb, which is very short, and armed with a strong curved claw.

Not only are the bones of the fingers elongated, but those of the palm of the hand, or ' metacarpals,' are drawn out to an astonishing length, that of the thumb being excepted, as above mentioned.

If we spread our own fingers widely, we shall see that their bases are connected by a fold of skin which is hardly perceptible when the hand is closed. Now and then, it is extended as far as the first joint, and there are many of the mammalia in which it is still further developed. The seals which fly through the water have the hand membrane greatly extended, and other water-living mammalia have it developed in a lesser degree.

Then, the flattened skin-fold of the flanks is not peculiar to the bat tribe. It can be traced in the common squirrel, and in the flying lemurs, flying squirrels, and flying opossums a similar structure is seen.

In the lower part of the arm, that from the elbow

to the wrist, there is practically one bone only. instead
of two, as is usually the case, the reason of which is
very apparent. It is owing to the two bones of the
arm that we are enabled to turn the limb inwards and
outwards at the elbow. If the bat were possessed of
the same power, it would be impossible for the wing
to strike the air with the steady beat necessary to
flight, for the resistance of the air, turning the arm
sideways, would allow the wing to cleave through it
sideways, and the power of the stroke would thus be
of no avail. As it is, however, being one bone only,
the bat is unable to turn the limb, which therefore
always presents its full surface to the air.

The bones of the hand, too, cannot be clenched as
in a fist, but possess a side motion only, enabling
the wing, when not in use, to be folded closely against
the body.

The membrane which forms the wing is merely a
prolongation of the skin of the flanks and other
parts of the body, stretched tightly between the
finger-bones, and extending as far as or farther than the
tail, which, in the insect-eating species, is included in
it, serving, like the tails of birds, as a natural rudder
by which the animal can direct its course. In the
fruit-eating bats, however, where so great agility in
the air is not necessary, the tail is left partly or
entirely free, and is much used in climbing and
walking.

The membrane is a double one, very thick in those
parts contiguous to the body, but so delicate near the
edges, that by the aid of a microscope the blood
corpuscles can be seen passing along the vessels that
supply the wing.

Though the powers of flight of the British bat are
fully equal to those of many birds, they have never been
known to migrate from one country to another, and

it is very doubtful whether they possess the ability. For bats have none of the large auxiliary air-cells found in birds, acting as a sort of reservoir, and their bones are not permeated with air-cells as are those of the feathered migrant.

There are, however, several bats belonging to the genus *Mycteris*, found in Africa, which possess a somewhat similar apparatus, though not constructed upon quite the same principles.

The skin is very loosely fastened to the body, a few membranous threads being the only bonds. The space between this loose skin and the body is utilized as an air reservoir, and is filled as follows. At the bottom of the cheek-pouches on either side is found a small opening, which can be closed at the will of the animal, and the air prevented from escaping. When the bat wishes to inflate its body, it closes the mouth, and forces the air from the lungs through the cheek-passages into the vacant space. To such an extent does it inflate itself, that it loses all resemblance to a bat, and looks merely like a round ball of fur provided with head and limbs.

The objects of this curious structure are not known, besides the evident one of increasing the buoyancy of the animal.

The shoulder-blades of the bat are enormously large, almost covering the whole of the ribs. These also are large and strong, and the breast-bone, besides being of unusual length, is furnished with a

Keeled Breast-bone of Bat.

C

central ridge, or keel, like that of birds, for the better attachment of the powerful muscles which work the wings. The rest of the skeleton is of the very slightest description, in order that no unnecessary weight shall hamper the movements.

The feet are very small in proportion to the rest of the body, and are furnished with long curved claws, which are of assistance in walking, but are chiefly used in assuming the extraordinary position of rest, when the bat hangs head downwards, from some convenient ledge or beam, merely hooking itself on by means of the claws. In this strange and, one would think, extremely uncomfortable attitude the bat always rests, and passes the winter in a torpid condition.

It was remarked many years ago, that the bat possessed a most wonderful power of avoiding any obstacles that presented themselves in its path, and that it could pass among the branches of trees, even where the twigs were thickest, without coming in contact with them. In order to ascertain whether this was always the case, a number of strings were stretched in a darkened place, and several bats let loose among them; yet it was found that the animals avoided them with the greatest ease. Thinking that this power might be the result of an unusually keen eyesight, one investigator, named Spallanzani, in a very cruel experiment, put out the eyes of a bat, and again let it fly, but was surprised to see that the creature avoided the objects exactly as before. It was then thought for many years that the bat possessed a sixth sense unknown to man, and it was not until comparatively lately that the true secret was discovered.

A careful examination of the membranes of the wings and ears showed that they were intersected by exceedingly delicate nerves, and it was found that

Bats Disturbed in their Cave.

the bat was thus made aware of the neighbourhood of an obstacle, and enabled to avoid it accordingly.

The fur of the bat is of a very soft and silky nature, and the hair is a most beautiful object under the microscope. It is densely clothed with scales somewhat resembling those of a butterfly's wing, which in some species are arranged in circles round the hair, a short distance from each other. The whole object bears a wonderfully strong resemblance to the well-known mare's-tail plant.

Easy and graceful as are the movements of the bat whilst disporting itself in the air, it is a very different creature when attempting to walk upon a level surface. Its mode of progression can at best only be described as an awkward waddle, the creature hitching itself along by means of the claw at the extremity of one of the wings, giving a kind of tumble forwards, at the same time advancing the corresponding foot; the same process is then repeated with the other wing.

In the illustration representing the bats in a cave, the extraordinary attitude assumed in walking is well shown. The long finger-joints are pressed together, their tips projecting on either side of the back. The weight of the body rests on the wrist, and the creature pulls itself forwards by hitching the claw of the thumb upon any roughness of the surface on which it walks.

Bats are remarkably averse to taking to the wing from a level surface, and always prefer to climb to some little height from which they can throw themselves into the air. This is evidently the reason for the strange and apparently uncomfortable attitude adopted when at rest, the animal being then in the most convenient position for launching itself into the air should there be any signs of danger.

It has been sometimes said that the bat is unable

Great Bat, Wing Closed and Open.

to rise from the ground, but such is not the case; and should the creature be hard pressed, it does not hesitate to do so.

It is able to climb with tolerable facility, and always does so with the tail uppermost, making its way up by the aid of the hind feet, the long claws of which are inserted into any convenient crevice in order to gain a foothold.

The food of all the British bats consists of the various small insects which fly about dusk, and at that time people the air in myriads. The appetite of the animals is almost insatiable, as may be gathered from the fact that a specimen of the short-eared bat, lately kept in captivity by ourselves, consumed daily from forty to fifty blue-bottle flies of the very largest dimensions, rejecting only the wings, and in a few cases the legs. Even upon this allowance, which seldom occupied it for more than twenty minutes, it did not thrive, but gradually wasted away, and finally died.

The bat was one which had been found in a hollow tree, and suffering from an injury to one of the wings, which entirely prevented it from flying. It was kept under a glass shade, into which the blue-bottles were introduced. It never took the slightest notice of the insects until nearly dusk, allowing them to crawl over all parts of its body without manifesting the least signs of activity. As soon, however, as the day began to close in, it was on the alert, and immediately set to work devouring the flies which had been procured for it.

This it did in the following manner :—

Resting upon the floor of its cage, it remained motionless until a fly settled within a few inches. It then began, by an almost imperceptible movement, to approach the insect, and when within an inch or so of it, with a sudden spring, clutched the insect between the wings, and holding them tightly together, bent down its head, and swallowed its captive. If a fly happened to take flight before the bat was near enough to make its spring, it merely remained motionless until another presented itself.

It was fiercely voracious when it once began to feed, and scarcely had one fly been swallowed than the bat was eagerly looking out for another. Its attitude when thus engaged strongly reminded us of that of the toad or the green crab when hunting after prey.

Though as a rule a nocturnal creature, owing to the habits of its prey, the bat may occasionally be seen flying in broad daylight, and sometimes, in the early spring, even hawking for the insects which are enjoying the warmth of the sun. In these cases, it is probable that the bat, having for the first time left the retreat where it had passed the winter in a torpid condition, has felt the want of food, and knowing instinctively that no insects would be on the wing at sunset so early in the year, has so far altered its usual habits as to prosecute its search by day instead of by night.

Most bats, however, resort to dark and retired hiding-places during the day, and in some parts of the world there are large caves which are celebrated as haunts of the bats. When travellers visit these caves, the guide will fire a gun into the cave for the purpose of startling the bats, which come rushing out in such numbers that unwary visitors have been fairly knocked down by them.

In Great Britain alone, there are nineteen catalogued species of bats, many of which, however, are rare, and very seldom seen. One of the commonest is the Long-eared Bat (*Plecotus communis*), which abounds throughout the British Islands. It derives its popular title from the great length of its ears, which stand out for some distance from the head, and which are thrown at every moment into a variety of graceful folds.   In consequence of its gentle temper, this bat is easily tamed, and is often kept as a pet, coming when called by those with whom it is familiar.

Another of the British bats is the Noctule, or Great Bat (*Noctulinia altivolans*), which is remarkable for the great height at which it usually flies. The specific name, *altivolans*, refers to this habit. It is the largest but one of the bats found in this country, being almost three inches in length from head to tail, while the spread of the wings is nearly fourteen inches.   Its cry is remarkably keen, and, like that of some other species, closely resembles the squeak which can be produced by rubbing two keys sharply together.   So shrill is the cry of the bat, that to many, among them even practised musicians, it is perfectly inaudible, the note produced being too attenuated to make any impression upon the ear.

A curious development of the nasal organ is found in the Horse-shoe Bat (*Rhinolophus Ferrum-equinum*), which is also a native of this country. It consists of a membrane, commencing at the lips, surrounding the nose, and projecting upwards for some little distance.   Immediately behind it is a second membrane, placed on the forehead, and sharply

Head of Horse-shoe Bat.

pointed.   It has been thought that the object of the structure is to increase the delicacy of the sense of smell. The same apparatus is found in the Vampire Bats, and is there developed to even a greater extent.

In other parts of the world, and especially in tropical climates, are found many other species of bat, some of them reaching the tremendous dimensions of nearly five feet in stretch of wing. This is the case in the well-known Kalong of Java (*Pteropus rubricollis*), which is often known as the Flying Fox, or Roussette. This, instead of feeding upon animal food, finds its subsistence in fruit, and is often the cause of terrible damage to the agriculturist. In some districts, indeed, where the bat is more than usually abundant, it is even necessary to envelope the whole of the fruit in a bamboo network, in order to secure it for human consumption. By way of a counterbalancing advantage, however, the flesh of the Kalong is eaten in many places, and is even considered a great delicacy.

A curious point about the Kalongs is, that they do not trouble themselves to find a dark and retired spot in which to pass the daytime, but hang in large clusters from the boughs of various trees, especially those of the fig tribe, where they are hardly recognisable as bats, resembling clusters of fruit more than anything else.

Perhaps the most widely known and famous of the bat tribe is the Vampire Bat (*Vampyrus spectrum*) of South America. It is by no means one of the largest of the family, its total length being six or seven inches only, while the spread of wings is not more than a couple of feet.

The blood-sucking propensities of this animal are well known, men and animals alike suffering from its attacks.

Settling upon its victim when plunged in slumber, it perforates with its sharp teeth any exposed portion of the body, generally selecting a toe as the point of operations, when it attacks a human being, and then sucking the blood from the wound until it is thoroughly satiated, a condition which seldom ensues until a considerable quantity of blood has been abstracted. The bite causes no pain at the time, and very little afterwards, the only ill effects arising from the loss of so large a quantity of blood.

The Vampire seems rather capricious in its tastes, for while one person may suffer from its attacks night after night, another individual, reposing perhaps only two or three feet distant, may leave his feet uncovered with perfect impunity, the bats never attempting to interfere with them. The late Mr. Waterton was one of these fortunate individuals. When travelling in British Guiana, with the hope of ascertaining exactly the mode of the vampire's attack, and the effects of the bite, he slept for several months in an open loft, purposely leaving his feet exposed. Yet, though the bats were frequently seen hovering over his hammock, and a young Indian, who also slept there, was repeatedly bitten, he was never attacked.

Cattle are great victims to the ravages of the vampires, which often reduce them to a mere mass of skin and bone by the frequency of their attacks. The wound is usually inflicted upon the flanks of the animal, just where the teeth or feet of the victim cannot reach it; and in cases where there is pressure from harness or other causes, often leads to considerable damage.

None of our British species are capable of harm, the teeth being too small to produce any impression upon the skin. Among the lower classes, nevertheless, bats are held in the greatest dread, and many a countryman would as soon handle an enraged viper as one of these harmless little creatures.

No. IV.—THE MOLE.

ONE of the most useful, and, at the same time, one of the least appreciated of the whole animal creation, is found in the Mole (*Talpa Europœa*), which, in spite of the inestimable benefit it confers upon agriculturists, is ranked by them as one of their worst enemies, and persecuted accordingly.

We can hardly pass through a meadow in many parts of the country without noticing a number of upright sticks planted in various parts of the field, to many of which is suspended a mole which has been captured and slain by one of the professional trappers employed by the farmer. Now, if the farmer had taken the trouble to look below the surface, and traced the mole through its day's work, he would have found that, instead of damaging his fields and ruining his crops, the animal was in reality rendering him services which could hardly be procured for money, and that instead of untiring persecution and ruthless extermination, it ought to be protected and encouraged to the utmost of his power.

But, as with many other almost equally useful creatures, so it is with the mole, which, placed in the category of 'vermin,' falls under the ban of the farmer, and pays with its life the penalty of human ignorance.

To us who inhabit the upper world, it seems a strange and comfortless life, this of the mole, spent in its cold, dark tunnels beneath the earth. We can hardly conceive a more wretched existence than one passed in an underground dungeon, damp and cold, and never cheered by a ray of sunlight. Yet our pity would be wasted were we to bestow it upon the mole, for no animal is better suited to its own mode of life, or more carefully adapted to obtain enjoyment from its apparently unpleasant surroundings.

Common Mole, and White or Albino Variety.

Indeed, the mole is quite as unhappy when taken from the darkness, damp, and cold where it lives, as we should feel if placed in a subterranean dungeon and deprived of the light, dryness, and warmth which are essential to our comfort.

In every detail of its structure, we see how well the mole is suited to the life it has to lead.

Cylindrical in form, its pointed snout and powerful digging-claws enable it to burrow through the soil with wonderful rapidity, while every sense is modified to suit the circumstances of its existence.

Like those of all the rest of the *Insectivora*, or insect-devouring animals, to which group it belongs, the teeth of the mole are formed for biting and seizing prey alone, and not for masticating the food, a few sharp pecking bites being the only preparation for swallowing it. It is by this structure of the teeth, also, that these animals are enabled to grasp and retain the struggling prey.

In order to fit the mole for its burrowing life, the strength of the fore-parts is developed to a wonderful degree.

The bones of the fore-limbs are stout, ridged, and considerably bowed, always a sign of great strength. The shoulder-blade, in particular, is of extraordinary length, in comparison with those of other animals, even that of the tiger or lion, either of which creatures can strike an ox to the ground with a single blow of its paw, fading into comparative insignificance beside it. In fact, if the mole were enlarged to the size of the tiger, it would be quite as active, by far the stronger and more terrible animal of the two.

This extra length of shoulder-blade is necessary for the attachment of the large and powerful muscles which work the fore-limbs, and which, when the skin is stripped off, can be seen lying in thick masses, almost as hard and strong as so much steel wire.

With this wonderful development of muscular power, it is no wonder that the burrowing powers of the mole should be so great, seeing, besides, what efficient digging instruments it possesses in its fore-paws. These are set obliquely with the body, in order to secure a larger scope for their movements.

They are large and powerful, not covered with fur like the rest of the body, and are furnished with long, curved, and rather flattened claws, which can penetrate the hardest earth.

These paws occasionally fulfil other offices than those of digging, for the mole is by no means a bad swimmer, and is often known to cross brooks and small streams, using the paws as paddles.

It is scarcely possible to select two creatures which present stronger contrasts to each other in point of structure than the moles and the bats, both insectivorous, but the one formed for flying in the air, and the other for burrowing in the ground.

In both animals the hind-limbs are but little required, and therefore they are feeble and comparatively insignificant, the chief distinction being in the development of the fore-limbs.

Beginning with the shoulder-blade, or 'scapula,' we find it large in both animals, but differently developed, in the bats being wide, thin, and covering many of the ribs, reaching as far as the pelvis.

In the mole, the scapula is long, narrow, very strong, and projecting upwards so as to afford attachment for the powerful muscles of the arm. The other bones are shortened and thickened in order to carry the enormous digging claws, and are deeply ridged for the attachment of the tendons which work the joints. In the bats they are attenuated to the last degree, and only one of them is capable of bearing a slight, hooked claw, solely employed for dragging itself clumsily over a level surface.

Yet the bones are the same in each case, and when the creature is dissected, the exquisitely perfect adaptation of each bone to its own office cannot but excite our highest wonder and admiration.

In the accompanying illustration the skeleton of

the fore-limbs is given, so that the reader may
compare the bones with those of the bat, which have
already been described and figured.

The structure of the long and flexible snout is
well calculated to aid the mole both in digging and
also in ascertaining the nature of the soil in the
places which it selects for its burrows; for, though
sensitive and possessing a keen sense of touch, it is
of great service in shovelling out the loosened earth—
a task for which it is further suited in some species
by the possession of a small auxiliary bone in the
tip. This auxiliary bone is also found in the snout

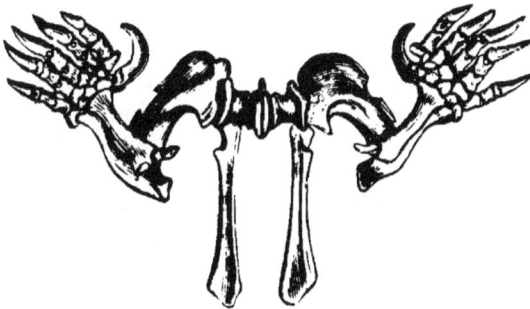

Common Mole. Bones of Fore-limbs.

of the pig, which in its wild state procures a considerable portion of its food by means of the digging
powers of that member.

While the animal is living, the snout is of a
pinkish colour, and of a very elastic nature; it is a
curious fact that, after death, it becomes hard and
wrinkled, and not even the most experienced taxi-
dermist can restore it to its pristine appearance.

Even the fur of the mole is arranged in such a
manner as not to impede its progress while passing
either backwards or forwards along its burrow. In

order to attain this result, each hair is finest nearest the body, gradually increasing in thickness towards the tip, and is set perfectly upright, in order that its resistance may not impede the animal in its movements.

When washed perfectly clean, and viewed in a good light, the fur is seen to be beautifully iridescent, all the colours of the rainbow playing over it in succession. This is still more strongly the case in the changeable mole of South Africa (*Chrysochloris holosericea*), which will be presently mentioned.

Attempts have been made to put the skin of the mole to various uses, only one of which, however, seems to have met with any particular success. This solitary instance is in the manufacture of purses, a custom much in vogue among the peasantry in some parts of the country. The operation is simplicity itself, the head and legs of the animal being taken off, the skin of the rest of the body dried, and the bag thus formed being closed by a string round the neck. In Wiltshire these purses are in very common use. From the great warmth of the fur of the mole, it has often been thought that it might be utilized in the manufacture of garments; and the late Mr. Frank Buckland, always fond of trying experiments in anything relating to natural history, procured a number of the skins, and had a waistcoat made from them.

However, he was never able to wear it, for two reasons—the first being that it was even *too* warm, so much so as to be almost unendurable; and the second, that it was impossible to get rid of the unpleasant odour which characterizes the mole, and which persistently clung to the skins for years after they had been separated from the bodies of their owners.

So strong is this odour, and so long does it retain

its power after the death of the animal, that the professional mole-catchers are accustomed, before setting each trap, to rub their hands with the body of one of their victims, which they carry with them for that purpose, in order that the wary animals may not detect the human scent about the trap, and be warned of their danger accordingly.

With the single exception of sight, the senses of the mole are developed to a very considerable extent.

Its hearing is proverbially acute. For instance, in Shakespeare's play of *The Tempest*, the deformed slave Caliban advises his friends, when they are about to rob Prospero: 'Pray you, tread softly, that the blind mole may not hear a footfall,'—a phrase which has since become almost a household word.

It is not to be supposed that this sharpness of hearing is entirely owing to the delicate structure of the mole's ear; for if it were to live in the open air, it is doubtful whether it would be able to hear better than any other animal. The fact is, that the earth is a very good conductor of sound, as may be easily proved by laying the ear upon the surface of a high road, when the noise of an approaching carriage may be distinctly heard while it is yet two or three miles distant.

Here, again, we have an instance in Shakespeare. In *Romeo and Juliet*, Act v. Scene iii., occur the lines :

'Under yon yew-tree lay thee all along,
Holding thine ear close to the hollow ground ;
So shall no foot upon the churchyard tread,
Being loose, unfirm, with digging up of graves,
But thou shalt hear it.'

The sense, too, of scent is particularly strong, enabling the mole to detect the presence of the insects and worms upon which it preys.

D

The sense of touch, also, is highly developed, more especially in the snout, which the animal uses to examine the nature of the soil into which it intends to burrow. If a mole on the surface of the ground wishes to sink a fresh tunnel, it may often be seen running to and fro, and trying various places with the snout, until it has settled upon one to its liking.

Although the senses of scent, touch, and hearing are so extremely sensitive, that of sight is little more than rudimentary. On casually examining a mole, the observer would be unable to detect the presence of eyes, which are deeply buried in the fur, whence, however, they can be protruded at the will of the owner. The creature can be forced to expose them by suddenly dipping it into a pail of water, when the mole, alarmed at the unexpected immersion, instinctively protrudes them from the mass of fur with which they are usually covered, looking like very small black beads. Even when thus exposed, the vision is very imperfect, and is, indeed, hardly necessary in the subterranean existence which the animal leads.

The FOOD of the mole consists chiefly of worms, grubs, and other small creatures which it finds beneath the surface of the earth. By the mere destruction of the dreaded 'wireworm' grubs, and the larvæ of the common cockchafer, which so often devastate the crops, it renders no small service to the farmer, and this alone should protect it from the persecution to which it is so constantly subjected.

Although it finds the greater part of its food in these creatures, it by no means despises prey of a larger nature, and will eagerly devour any small bird or mouse which it may happen to meet with.

Its voracity is something extraordinary; for it kept as a pet, it is one man's work to keep it supplied

with worms. It is almost impossible to describe the
ferocity with which it devours its prey.

As soon as a worm is put into the cage, it detects
the presence of food as if by magic, springs upon its
victim, and rapidly forces it into its mouth with the
fore-paws, giving meanwhile a series of rapid crunch-
ing bites, and causing the unfortunate worm to disappear with marvellous celerity.

No sooner is the first worm swallowed than it is
on the look-out for a second, which is speedily disposed of in a like manner. So voracious is its appetite,
that it is said to be unable to endure a fast of more
than three hours. Among the peasantry it is commonly
reported that the animal alternately works for three
hours and sleeps for three hours, and this seems very
likely to be the case.

The mole seems to suffer greatly from thirst, and
always digs a series of wells in different parts of his
burrows, to which it can repair when in need of
moisture.

The passions of the mole are all of the fiercest
nature, and when enraged, it seems utterly devoid of
fear, attacking an enemy far superior in size to itself,
and fighting with the greatest ferocity until death
puts an end to the scene. One mole was even
known to turn upon the individual who was holding
it, and inflict a severe wound, refusing to quit its
hold until almost killed by the teeth of its victim, no
other means proving of avail.

If, by any chance, two strange moles should happen
to meet in a burrow, there is but one invariable ter-
mination to their encounter. They fight, and the
conqueror devours its vanquished foe.

The rather unsightly although useful 'molehills,'
which are so plentiful in many places, serve to show
the course which has been followed by the mole, the

hillocks being merely the superabundant earth which it is obliged to throw out at short intervals. These tunnels mostly terminate in the 'fortress,' as it is termed, which is a really fine specimen of excavation, almost attaining the rank of architecture.

The situation selected is generally at the roots of a tree or large bush, where the ground is unlikely to give way above it.

The general plan of the structure is as follows :—

In the centre is a rather large circular chamber with exits at various places, which lead into a gallery surrounding it. Above this is a second circular gallery, communicating with the lower one by no less than five passages.

A large passage opening into the high road, is driven down from the lower gallery, and a large series of tunnels, radiating on all sides, and all communicating with the lower gallery, are finally constructed.

It will thus be seen that the mole, if chased by an enemy able to follow it along the tunnels, can take refuge in the fortress, pass through the centre chamber, and escape by one of the passages on the opposite side, leaving its foe bewildered in the complicated maze of tunnels. The reader must not, however, think that every fortress contains the whole of these passages and galleries ; and, indeed, it is very doubtful whether any individual fortress contains both the circular galleries and all the connecting passages.

The central cavity is usually filled with a quantity of moss, dead leaves, grass, etc., and is used as a bed-chamber by the mole during the colder seasons of the year. In warmer weather, however, it generally takes up its abode in one of the ordinary hillocks.

Generally at some little distance from the fortress, the female mole constructs her nest, which she builds in some large hillock, and lines with moss or dried grass. In this she brings up her young, usually from four to six in number, and provides for them until they are able to take care of themselves. Some good examples of the mole's nest may be seen in the museum at Liverpool.

The colour of the mole is usually of a blackish grey, somewhat paler upon the under side, although it varies to a considerable extent. Some specimens have been found of a pure white, and pale varieties are by no means uncommonly taken.

Putting on one side the immense benefit conferred upon the farmer by the wholesale destruction of 'wireworms,' the larvæ of the cockchafer and daddy-long-legs, which, feeding upon the roots of the crops, cause wholesale devastation, the mole is of the greatest service to the agriculturist in more ways than one.

His complicated network of subterranean passages, daily and hourly extended, not only forms a nearly perfect system of subsoil drainage, which could with difficulty be equalled by human labour, but the fresh earth brought continually to the surface from a considerable depth below the reach of the plough or spade, acts almost like manure in increasing the fertility of the land, and renders it capable of nourishing the crops with which it is planted. All that is necessary in a mole-inhabited meadow is to apply a rake to the heaps of earth, and spread them evenly over all parts of the field, in order that every yard shall receive its due share.

In yet another way does the mole prove himself the friend of the farmer, for by means of the loosened earth the air is enabled to reach the roots of the plants, where it is so much needed, and where it would have little chance of reaching were it not for the beneficent and untiring labours of the mole.

With such claims upon the farmer, it seems strange

that the habits of the animal should be so little understood and appreciated, and we can only hope for the time when the spread of zoological knowledge shall have shown that both this, and many others of the so-called 'vermin,' instead of being persecuted, should be protected and encouraged to the utmost of our power.

TURNING to foreign countries, we find several very near relations of our common mole, some of which present very great peculiarities both in habits and form. One of the most remarkable of these animals is the Chrysochlore, Shining Mole, or Changeable Mole (*Chrysochloris holosericea*), found chiefly in the Cape of Good Hope. As before mentioned, the fur of this creature possesses a brilliant metallic radiance, changing in various lights, and far superior to that of our British example, beautiful though that is. The scientific title is singularly appropriate, the name *Chrysochloris* being formed from two Greek words signifying gold-green; and *holosericea, i.e.* wholly silken, referring to the texture of the hairs.

In other ways, also, the Chrysochlore is worthy of notice. The digging paws are formed after a very singular fashion, being provided with four toes, the last of which is but of small size. The remaining three, however, are furnished with very long and powerful claws, the middle one especially being of surprising dimensions.

The jaws of this species are constructed after a perfectly unique fashion, a gap equal to the width of a tooth being left between each; so that when the jaws are closed, the teeth of each jaw fit into the interstices of the opposite one like those of a steel trap.

In the skeleton, also, are found several peculiarities.

one being that there are no less than nineteen pairs of ribs. The tail is entirely wanting.

Another of the foreign moles, and one of a very extraordinary appearance, is the Radiated or Star nosed Mole (*Astromyces cristatus*), sometimes known as the Condylure, which is found in Canada and the United States.

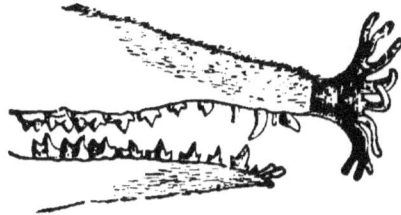

In this animal the tip of the snout

Snout of Condylure.

is modified into a number of pink, fleshy rays, branching off in every direction, and sometimes being as many as twenty in number. These rays are retractile at will, and are supposed to aid the animal in its delicate sense of touch, and in procuring the worms, etc., on which it feeds. Another curious point about the Condylure is the size of its tail, which sometimes exceeds two and a half inches in length. The name Condylure is formed from two Greek words, the former signifying a knob and the other a tail. It was given to the animal by a person who had only seen the dry skin and not the living creature. Except for its great comparative length, the tail has little about it that is remarkable. But, when the animal is dead, the skin contracts so forcibly over the vertebræ of the tail, that the organ looks something like a row of roundish beads strung upon wire and covered with skin.

No. V.—THE CAT TRIBE.

Part I.

INCLUDED among the members of the cat tribe—or *Felidæ*, as they are scientifically termed—we find many of the largest and most powerful of all the *Carnivora*, or flesh-eating animals. The lordly lion, the fierce and savage tiger, the crafty leopard, and many others, all belong to this family, which, Australia excepted, is spread over the greater part of the world

Intended by nature for an active and predacious existence, the structure of the Cats is pre-eminently adapted to suit their mode of life, and the whole form is a marvellous combination of strength, lightness, and activity.

Take the lion, for example. Who would think that an animal of such ponderous size and weight, who can strike an ox to the ground with a single blow of his mighty paw, could glide noiselessly through the thickest jungle and overtake an antelope in fair chase?

Let us now proceed to examine this wonderful structure in detail, and afterwards to devote a short space to each of the more prominent animals of the group.

We will first examine the skeleton, beginning with the skull.

The first point which strikes the attention is the bony ridge which runs along the crest of the skull, and which chiefly serves as the attachment for the powerful muscles which act upon the jaws. Inside the skull a

Skeleton of Cat.

second ridge of bone is found, partly separating the two great divisions of the brain from one another. The object of this ridge has never been satisfactorily proved, but it is thought to be of service in guarding the brain from the severe shocks to which it might otherwise be subjected from the leaps and bounds of the animal.

The teeth of the cats, like those of all other exclusively flesh-eating animals, are formed for biting and tearing the prey alone, and not for masticating it before it is swallowed. None of them are found with the flat surface necessary for grinding the food, and, even were such teeth possessed, the construction of the jaws would render impossible the side motion necessary for mastication. Upon watching a cat devouring her food, it will be seen that she bolts it in large lumps, swallowing it by a succession of sharp, pecking bites.

Passing to other parts of the skeleton, it will be noticed that the two first vertebræ are provided with a similar enlargement to that of the skull, for the better attachment of the muscles.

The bones of the limbs are of great strength, as is necessary for the rapid and powerful motions of the animals. The muscles are particularly hard and tough, seeming almost like bands of iron, and turning the edge of the sharpest knives.

The claws, which are the chief weapons of the animal, are long, sharp, and strongly curved.

All the animals of the tribe being *digitigrades*, or those which walk upon the tips of the toes alone, it might be thought that the points of the claws would be constantly worn down by the friction with the ground, and thus rendered useless for purposes of offence.

In order to prevent this, there is a beautiful mechanism which, when the paw rests upon the ground, withdraws the claws into the sheaths provided for them, and protects them from injury by the ground ;

Claw of Lion, sheathed.

as soon, however, as the paw is thrust forward to strike or grasp the prey, they are mechanically thrown out ready for use, the tendons being stretched, and therefore pulling them forwards.

Claw of Lon, protruded.

Most people have had their hand licked by a favourite dog as a mark of affection, and know that his tongue is wet and smooth. Not so that of the cat, which is comparatively dry, and provided with a number of sharp, file-like points, pointing backwards, and serving to scrape every particle of flesh from the bones of a slaughtered animal.

So rough is the tongue of the domestic cat that it
will cause pain to a delicate skin, while the larger
animals of the tribe will draw blood almost immediately.

There is a well-known story of a gentleman who
owned a pet lion, of which he was very fond, and with
which he often indulged in a game of play. One day,
whilst lying asleep upon a couch, the lion came up,
and seeing one of his master's hands hanging outside
the covering, began to lick it, just as a pet dog would
under the circumstances.

In a short time the rough tongue cut through the
skin and drew blood, which was greedily licked up by
the animal, the pain at the same time awaking his
master.

No sooner did he attempt to withdraw his hand
than the creature uttered an angry growl, whereupon
the gentleman, knowing the danger in which he was
placed, and that a moment's hesitation might seal his
fate, drew with his other hand a loaded pistol from
beneath the pillow and shot his favourite through the
head, as his only chance of escape.

All the feet of the cats are provided with soft, fleshy
pads, which enable them to move noiselessly from
place to place, and also serve to break the fall from
the long leaps and bounds of the animals.

Another point to be noticed, too, is to be found
in the whiskers.

These are provided at their roots with an extremely
sensitive nerve, so that the slightest touch is at once
felt by the animal. Those springing from the sides
of the mouth, too, are of exactly the same width as the
body, so that the animal is able to tell, even in the
dark, whether it can pass through a narrow orifice or
not.

Having now noticed the chief points of the structure

of the cats, we will take each of the more important members of the family in their order, and examine their habits and mode of life.

To begin with, we will take the Lion (*Leo barbarus*), which stands at the head of the cat tribe—the acknowledged King of Beasts. No animal, and scarcely even man, can hear without trembling his mighty roar, and from his strength and courage he is feared by every denizen of the forest.

It is yet uncertain whether or not there is more than one species of lion. Some authors suppose the African lion, the Gambian lion, the Asiatic lion, etc., to be different animals; while others consider them to be merely varieties of the same type, slightly modified according to the country in which they live.

The most widely known of these species, or varieties, is the African lion, which is spread over the whole of the southern part of that continent, excepting those parts where civilised man has gained a permanent footing, and driven the wild beasts from his neighbourhood.

The lion, when it has spent its life free and untrammelled in its native haunts, attains to considerable dimensions, a full-grown animal averaging some four feet in height at the shoulder, and nearly eleven feet in total length from the nose to the tip of the tail. The lioness is rather less in size, and, owing to her want of mane, appears even smaller in comparison than is really the case.

The colour of the lion is a dark tawny yellow, deeper on the back, and lighter on the under parts of the body. The ears are blackish, and there is a thick tuft of hair at the end of the tail, found in no other member of the cat tribe, which is also black. The male lion, when it has attained the age of three or four years, is furnished with a shaggy mane of long hair, which falls from the neck and shoulders, and part of the throat. In the female this mane is wanting. It is remarkable that when young the lion bears dark stripes and spots on its fur, so that a well-marked specimen might easily be mistaken for a tiger-cub. A similar arrangement of colour is found in several animals, such as one or two species of swine, and the Malayan tapir.

There are various opinions as to the character of the lion.

Some hold him up as a model of generosity and courage, sparing the weak, but fighting the strong with the utmost courage. Other writers represent him as a mean, sneaking animal, afraid to face an armed man, and preferring to obtain his prey by stealth rather than by open warfare. Others, again, consider that his temper is uncertain, and that one day he will fight with great courage and ferocity, while on another he will fly in terror from a comparatively insignificant foe.

It is certain that the courage of even the same lion seems to vary at different times, and that on one occasion he will openly attack a number of armed men, while on another he takes refuge in flight before a single savage.

Like the rest of the cat tribe, the lion is of a very indolent nature except when suffering from the pangs of hunger, and, unless he be in want of a meal, will seldom take the trouble to fight.

Even when in pursuit of prey, he never takes more exertion than is absolutely necessary for his purpose, but prefers to stalk the quarry until he can creep within five or six yards. One powerful bound and a single blow of his paw then suffices to kill his prey, and he earns his meal with very little active exertion.

Not that he cannot overtake most animals if he is forced to do so, but, just as the cat prefers to spring unawares on a mouse rather than chase it, so does the lion act towards the larger animals on which he preys.

Should the animal be of such size that the blow from his paw would be insufficient to kill it, the lion springs upon the flank or shoulder, and drags it to the ground by sheer strength, when it is easily despatched. A lion and his mate have been seen to spring upon a giraffe, and by their combined efforts to tear it to the ground.

Although the lion prefers living animals for his prey, he by no means disdains a repast from any carcase which he may find lying in his path. It is thought by several writers that many " man-eaters " acquire their taste for human flesh by preying upon the bodies of slain natives which they find in the bush.

If large game be scarce, he will satisfy his hunger by a meal on some of the smaller rodents, and has even been known to devour locusts and other insects in times of scarcity. If he should happen to suffer from thirst, too, when water cannot be procured, his instinct teaches him to search for the juicy water-melons which grow in the desert, and which answer the purpose of liquid nourishment to many animals ; so that we have the remarkable fact of a carnivorous animal voluntarily taking to vegetable food.

The lion is justly dreaded by the colonists of Southern Africa, for hardly a more determined foe to the farms could be found. Night after night he visits the enclosures, carrying off a valuable animal at each visit, and making his raids with such cunning and ingenuity, that it is generally a most difficult matter to trap or shoot him. Favoured by the darkness of the night, he creeps close to the folds, carefully watching for every sign of danger ; as soon as he is satisfied that his presence is unnoticed, he leaps among the cattle, strikes one of the animals to the ground, and drags it off into the bush before the alarm is fairly given.

Travellers passing through the country with a train of horses and oxen, often lose a considerable quantity of stock by means of the lion's nocturnal visits. It is always the custom, at the approach of night, to tether the horses and oxen to the bushes, and making a large fire, to form an encampment around it.

The lion will prowl round and round within twenty or thirty yards of the camp, but dares not approach nearer on account of the fire. Finding that the cattle do not seem inclined to stray, he retires to a short distance, places his mouth close to the ground, and gives vent to two or three of his loudest roars. The oxen, alarmed by the terrible sounds, often break away from their tethers and rush out into the darkness, when the lion has no difficulty in making one of them his victim.

The most dreaded of all, however, is the terrible "man-eating" lion, which prowls in the neighbourhood of the villages, ready to pounce upon any unprotected human being who may pass within the vicinity of his lair. When once a lion has tasted human flesh, he prefers it to every other kind of food, and, daily growing bolder in his raids, causes a perfect panic in the neighbourhood. In such a case, the whole population of the village takes the field, and there is no rest until the dreaded man-eater is slain.

Several lions often band together in search of prey, and act in concert, each having his appointed part. In such a case, one of the lions drives the prey, generally a herd of elands, or other large animals, towards his companions, who lie in wait until the flying animals are within reach, when they fall an easy prey to their hidden foes.

Many stories of adventures with lions have been told, in some of which the lion has reversed the wished-for order of proceedings, and slain the hunter instead of being himself killed.

Even when mortally wounded, the animal generally has sufficient strength to throw himself upon his assailant, often killing or severely wounding him before succumbing in the death-struggle. Experienced hunters, therefore, always try to conceal themselves until the lion has been forced to yield himself to death. The few seconds of furious life that remain in the stricken animal may suffice to lay the hunter beside his dead enemy.

In a few instances it has happened that a hunter has been carried off by a lion, and has yet escaped with comparatively slight injuries. The lion, like most of the cat tribe, when he has captured any animal, prefers to play with it for a short time before killing and eating it, just as our domestic cat amuses herself with a mouse, sometimes for hours, before putting an end to its miseries.

Knowing this habit, the hunter has remained perfectly quiet until the lion placed him on the ground, when, drawing a pistol or knife, he has contrived to shoot or stab his foe to the heart, and thus rescue himself from a horrible death.

Sometimes, as in the well-known case of the late Dr. Livingstone, it has happened that the lion has dallied with its prey long enough to permit the comrades of the fallen man to come to his rescue.

Men who have escaped in this manner always say that the first shake of the lion deprived them of all sense of fear and pain, and that they were then only conscious of a kind of languor, mixed with a wonder as to the way in which the lion intended to eat them.

It has often been noticed, in the case of a mouse caught by a cat, that even when released by its enemy it did not seem to attempt to save itself, but moved as though fascinated and unable to fly.

So we may conclude that the same merciful provision is made for the various animals upon which the carnivora feed, and that after the first shock little or no pain is felt.

Livingstone, who was once carried off by a lion, compares the sensation to that of a patient, partially under the influence of chloroform, who sees all the operation, but feels not the knife. 'The shake,' he says, ' annihilated fear, and allowed no sense of horror in looking round at the beast. This peculiar state is probably produced in all animals killed by the carnivora ; and, if so, is a merciful provision by our benevolent Creator for lessening the pain of death.'

Even should a man succeed in making his escape from the clutches of a lion, and his wounds heal in due course, he is very unlikely to have entirely recovered from the effects of his injuries, for there seems to be a peculiar poisonous property about the lion's teeth and claws, causing the wounds made by them to break out afresh every year about the time at which they were first inflicted.

The lion is an exceedingly cautious and wary animal, and will carefully avoid any object which it does not understand, or which bears the least resemblance to a trap.

Knowing these cautious habits, the hunters are accustomed to protect the carcase of a slaughtered animal, which they are unable to carry away at the time, by fastening a streamer of white cloth to a stick, and planting it in the ground close by the animal. Or, a kind of clapper is constructed, which rattles in the wind, and which is planted in the same way. Although the lions will prowl in the neighbourhood all night, not one will dare to approach so mysterious an object, even though he may be half wild with hunger.

The young of the lion are generally from two to four in number, and are about the size of an ordinary tom-cat. For the first few months of their life, the fur is brindled with darker stripes, in the same way as that of the tiger ; as they grow older, however, these stripes gradually become fainter, and at last entirely disappear. The cubs are wonderfully playful little animals, and frisk and gambol about their mother just like so many kittens. Their weight is very great in proportion to their size, the skeleton and muscular systems being very solid and massive in structure. The full growth of the lion is not reached until the end of the fourth year.

In the more northern parts of Africa a variety, or species, of the lion is found, usually known as the Gambian lion (*Leo Gambianus*). In character and habits it differs only in the very slightest degree from its southern relative.

The lion which inhabits Asia presents no particular points of difference from that of Southern Africa, and a detailed description will therefore be unnecessary. A peculiar variety, or species, according to some writers, however, is found in Guzerat. This is usually known as the Maneless Lion (*Leo Goojrattensis*), on account of its lacking the hairy covering of the neck and shoulders which is found in the other species. This is not entirely wanting, but is very imperfectly developed, the animal not possessing, in consequence, the majestic aspect with which we are so familiar. The tail is shorter in comparison, with a large tuft of hair at the tip.

By the natives, this animal is often called the 'camel-tiger,' on account of the resemblance its fur bears in colour to that of the camel.

Like most of the members of the cat tribe, the lion has often been tamed as a pet, and learned to follow and obey its master just as does a dog. In these cases the animal has always been captured while quite a cub, before the wild and savage instincts of its nature had yet been implanted in its breast. It is commonly the case, too, to find a troop of performing lions in menageries and circuses. No matter how well trained the animals may be, however, it is always a dangerous performance, for a slight whim or a piece of ill-temper on the part of one of the animals may bring out all the savage nature of the beasts, and, before aid can be given, cause the death of the trainer, as has but too often happened.

No. VI.—THE CAT TRIBE.

PART II.

NEXT in order comes the Tiger, which is spread over a considerable part of Asia. The animal which is popularly called the 'tiger' by African hunters is only a large leopard, and the 'tiger' of American hunters is the jaguar. It is by no means equally distributed, some parts of the country being absolutely infested by the animals, while in others they are seldom or never seen.

The tiger is fully equal both in size and strength to the lion, and certainly surpasses that animal in the ease and grace of its movements. Its colour is a bright, tawny yellow, with a number of dark brownish-black stripes, some of which are double, running round the body at right angles to the limbs. On the lower parts of the body the fur becomes nearly white, and the dark stripes melt almost imperceptibly into the general ground colour.

Occasionally a tiger is found whose fur is of a uniform greyish-white hue, the stripes being scarcely visible. This is usually known by the name of White Tiger, but is merely a variety of the common species.

It seems strange that so brightly-coloured an animal as is the Tiger should be so extremely inconspicuous among the underwood of its native jungles. Such, however, is the case, and a tiger at the distance of ten or fifteen yards would be perfectly invisible except to the most practised eye. For the dark stripes harmonise so perfectly with the dark shadows between the upright blades of the long grass, and the tawny yellow fur so strongly resembles the bright hues of the surrounding foliage, that, until it moved, even the most experienced hunter would probably be unaware of the presence of the animal.

Like the lion, the tiger seldom undertakes more active exertion when in pursuit of prey than is absolutely necessary for the attainment of his object. He seldom or never attempts open chase, but prefers to stalk his quarry, sometimes for miles, gradually creeping closer and closer, until he is able to effect his purpose by means of a single bound.

In the same way the dreaded 'man-eaters' will follow human beings, generally devoting their attention to women and children, who are not likely to carry weapons.

The mortality from these animals is very great, for a tiger, when once it has tasted human blood, ever after thirsts for it, just as is the case with the lion. In some districts, even, a victim is carried off almost daily, the mingled apathy and superstition of the natives allowing the animal to carry on his depredations with impunity.

The spots most infested by tigers are those localities where the road passes through a small copse, or patch of jungle, and where water is in the immediate neighbourhood. Where water is scarce, in fact, the tiger is seldom found, as he requires to quench his thirst after every meal.

In such a retreat he lies in wait, always upon the opposite side of the road to that on which his lair is situated, until some unfortunate animal, or human being, happens to pass by. With a tremendous spring he bounds upon his victim, dashes him to the ground, and drags his body across into his lair without being obliged to turn round.

Should he happen to miss his aim, as does sometimes happen, he seldom or never repeats his attack,
but seems bewildered, and mostly slinks away among
the bushes. Should a number of people be together,
he always selects the last of them, so that in tiger-
hunting the post of honour is, as in a retreat, in the
rear.

The natives divide tigers into three kinds, namely,
the Hunting tigers, the Cattle-eaters, and the Man-
eaters.

The first are the younger animals, which are strong
and active enough to hunt prey for themselves. The
natives do not try to destroy these animals, and find
them rather beneficial than otherwise, because they
keep down the antelope herds that make havoc in
the grain-fields.

The second are the older animals, which can no
longer trust to the chase for food, but hang about
villages for the purpose of pouncing upon any stray
cattle that may come in their way.

No less than seven such tigers have been driven out
of one cover, so that the destruction which they work
can easily be imagined.

Their mode of attack is always the same. They do
not knock down their prey with a blow from the paw,
as is generally imagined, but seize it by the nape of
the neck, and with both paws on the head, twist its
neck.

A single tiger has been known to destroy annually
between sixty and seventy head of cattle, none cost-
ing less than five pounds, and many being worth
double the money. These cattle-eaters are curiously
fastidious. When they have killed an ox, they drag it
to their feeding-place, and then open and clean the
body as neatly as any butcher could do, always putting
the offal at some distance from the meat.

Old tigers, who cannot even destroy cattle, are tolerably sure to become man-eaters. Fortunately for the natives, European huntsmen never allow a man-eater to live. It is even necessary to destroy every cub of a man-eater, for if a tiger, no matter how young, has once tasted human flesh, it becomes at once a man-eater.

The fur of such an animal is never worth anything in a pecuniary point of view, as it is almost always mangy, bald in patches, dingy in hue, and never exhibits the rich, warm colouring of the healthy fur.

So great is the terror of the natives at the mere presence of a man-eater, that they seem quite demoralized. They never venture out at night, and even by day will only dare to move in large bodies, all being heavily armed and accompanied by the beating of drums and the shouts of men, the firing of shots, and the glare of flaming torches.

A single man-eater has been known to kill a hundred human beings in a single year, and to put a stop to traffic in a triangular district measuring from thirty to forty miles on each side. Many villages were wholly deserted, and others in which the inhabitants remained, were surrounded with strong palisades.

The claws of the tiger form most terrible weapons, being sickle-shaped, and as sharp as a knife. As is the case in the lion, the claws seem to possess some poisonous influence apart from the actual wound, for, in many cases, even a slight scratch has been productive of lockjaw, followed rapidly by death. One hunter, of many years' experience, states that he has never known a patient to die from the effects of a wound caused by the tiger's claws without suffering from lockjaw previous to death.

Naturally, no pains are spared to exterminate so

powerful and dangerous an animal, and traps of all kinds are constructed for his capture.

Some of these are most ingenious. A very common method of destruction is the spring bow, which is set as follows :

Two stout posts are planted in the ground by the side of the tiger's path, and to these the bow is firmly fastened, the string being parallel with the path. The bow is then stretched, and kept in that position by means of a stick, which prevents the string from approaching the wood. At the end of the stick is placed a long wedge, to which is fastened a cord, which crosses the path of the animal. The arrow, generally poisoned, is then laid in its place.

Naturally, as soon as the tiger presses the cord with his breast, the wedge falls, the stick is drawn away, and the arrow discharged into his body, where the poison very shortly proves fatal.

Should a tiger have paid a visit to a farmyard, and carried off a horse or bullock, his track is followed up until the carcase of the slaughtered animal is dis covered. Knowing that the tiger will shortly return for a second meal, the farmer cuts a few gashes in the flesh and introduces a quantity of arsenic. Before very long, the tiger makes his appearance, and swallows great lumps of the poisoned food, which in a short time puts an end to his existence.

Farmers in this country are familiar with a method of catching rooks, when they attack the newly-planted grain, by twisting up a number of paper cones, placing a grain or two of corn at the bottom, and smearing the interior with bird-lime. These are placed in the furrows, with the pointed end downwards. The rook comes flapping along, sees the corn at the bottom of the cone, and immediately attempts to secure it. His head once in, however, he is unable to release it, the tenacious bird-lime fixing the cone over his head, and preventing him from seeing. After a short time he is exhausted by his struggles, and is then easily secured.

In much the same manner tigers are often captured.

A number of the broad leaves of the *prauss*-tree are secured, and thickly smeared with bird-lime. These are laid in the animal's path, the hunter concealing himself in the neighbourhood.

The tiger passes along, and treads upon one of the prepared leaves, which adheres to his foot. Not being able to remove it, he rubs his paw against his head, after the fashion of the cats, thereby transferring the sticky substance to his ears and eyes. By this time he has trodden upon more leaves, which serve to still further incommode him, and he struggles to free himself from the mysterious substance, rolling upon the ground in his efforts, until he has completely covered himself with the bird-lime. Guided by his voice and struggles, the hunters come up and despatch him without difficulty.

Sometimes a building resembling a huge mousetrap is constructed, and baited with a sheep or goat, which is placed in an inner chamber, so that it cannot be reached from the outside by the claws of the tiger ; or a large bamboo cage is built, the hunter taking up his position inside, and spearing his foe through the bars as he ventures to attack.

Sometimes a large bamboo platform is erected near the haunts of the animal, on the summit of which the hunter takes up his station, firing at the creature the moment it appears. Even should the wound not prove instantaneously fatal, and the tiger attack him, he is in perfect safety, being above the reach of its claws, while the polished bamboo affords no foothold to his infuriated enemy, who is easily killed by a second shot.

When a hunter has been fortunate enough to kill a

tiger, he always preserves the teeth and claws as tokens
of his success, and the natives would not think of
leaving the dead tiger without burning off its whiskers,
as a kind of charm.

Besides these manifold traps, the tiger is also hunted
in various ways, the most usual being by means of
elephants. Upon these animals ride the hunters, who
are seated in the 'howdah' (pronounced 'hōodâh'),
a sort of open carriage firmly fastened upon the back
of the elephant. A large number of beaters are pressed
into the service, who endeavour, by means of shouting,
blowing horns, beating drums, letting off fireworks,
etc., to drive the tiger from its concealment.

In spite of the size and strength of the animals
ridden by the hunters, this sport is not without danger,
the tiger often facing his pursuers, leaping upon the
elephant, and even reaching the howdah.

It is only by careful training that the elephants are
induced to face the infuriated beast at all. First, they
are taught to familiarize themselves with a stuffed skin,
and to gore it with their tusks, and trample upon it.
Next, a boy is placed inside the skin, in order to
counterfeit the movements of the animal, and accustom the elephant to the sight of the skin in motion.
Finally, a dead tiger is shown to the animal, instead
of the stuffed skin.

Yet, with every precaution, and the most careful
training, even the most courageous elephant will sometimes turn and run before an angry tiger, in spite of
the exertions of the 'mahout,' or driver, who rides
upon the neck of the creature.

One would naturally think that so destructive an
animal would be almost universally sought after and
destroyed. Yet in many parts of the country the
tiger is absolutely protected, being considered as a
sacred animal, and treated accordingly. Many of the

native chiefs, too, protect it for hunting purposes, just as the fox is preserved in our own country.

The tiger is a good swimmer, and has even been known to board vessels lying at a considerable distance from the shore, causing the greatest consternation among the crew.

The young of the tiger are two or three in number, and do not arrive at their full growth until three or four years have passed.

Owing to the colouring of the skin, to which allusion has been made, the tiger can with difficulty be discovered, even when its haunts are known. Hunters say that a tiger can hide itself in places where a rat could hardly find cover.

Practised hunters are always on the look-out for indications of the tiger's presence, one of which is a bush covered with berries. If no tiger were hidden there, the monkeys would not have left a berry on the bush, but as from their strongholds in the treetops they can see the enemy, they take care to keep their distance, and so let the berries remain on the branches.

Peacocks, again, are mostly found in places where the tiger lives. The bodies and feathers of dead peafowl are sometimes found strewn about a tiger's den. The natives account for this fact by saying that the tigress teaches her growing cubs how to hunt prey for themselves, and that they practise on peafowl before they can aspire to antelopes or cattle.

As to the size of a full-grown tiger, it varies almost as much as does the height of man. The average length of an adult male tiger is about nine feet six inches, measured from the tip of the snout to the end of the tail. A ten-feet male is as unusual an exception to the ordinary dimensions of tigers as is a man six feet three inches in height among ourselves. Measurements of the skin after it is removed from the animal

The Tiger.

are quite fallacious, a skin being capable of almost any amount of extension by stretching. To be worth anything, the measurements should be made before, and not after the skin has been taken off.

It is a curious fact that the mother does not seem so careful for the welfare of her offspring as is usual among animals, but, if she suspect danger, will often send her cubs on first, in order to see whether the path be clear. Experienced hunters, aware of this, refrain from firing at the young, knowing that the mother is behind, and will soon make her appearance.

Next in order is the Leopard (*Leopardus varius*), which is found both in Asia and Africa. It is by no means as large and powerful as the tiger, but is even more graceful in its movements. The colour of its fur is a bright golden yellow, closely studded with rosette-shaped dark spots.

A few leopards have been occasionally found whose fur was so dark as to earn them the title of Black Leopards, which were for some time supposed to constitute a separate species. However, it was found that the dark spots were still dimly visible, and that, except in point of colour, there were no particular differences between these black leopards and the ordinary animal, and that therefore they could only be considered as a mere variety of the common species.

To the leopard belongs a power which is not possessed by the lion and tiger—namely, the ability to climb trees. So quick and agile are its movements among the branches that it is even able to chase and capture the various tree-frequenting animals in their native haunts.

In some ways the leopard is even more dreaded than its larger and more savage relatives, especially by t' e farmers, who suffer greatly from its depredations

among their flocks. Combined with great agility, he
possesses the craft and cunning of the fox, and, like
that animal, usually selects the hen-houses of the
neighbourhood for his nocturnal raids. In these he
commits the greatest havoc, striking the birds to the
ground before they are even aware of the presence of
their enemy, and following them into the trees should
they roost among the branches.

The mischief he commits is rendered even greater
by his custom of storing up provisions for a rainy day.
For this purpose he usually selects the junction of a
large branch with the tree-trunk, and constitutes this
his larder, which he carefully conceals by means of
dead leaves, etc. He has even been known to carry
the body of a slain child into the fork of a tree, and
hide it there.

When on the look-out for prey, the leopard generally
conceals himself among the branches of some tree beneath which game is likely to pass. From his leafy
retreat he can then leap down upon the unfortunate
animal, and bring it to the ground merely by the force
of his spring. When hunted, too, he almost always
takes refuge among the boughs of a tree, and displays
great sagacity in selecting a spot where he is protected
from the aim of his pursuers.

On ordinary occasions the leopard is a much more
timid animal than most of his relatives, and is easily
frightened if taken by surprise. When driven to bay,
however, he fights with the greatest ferocity and des-
peration, dashing savagely at his foes, and wreaking
his vengeance upon them with tooth and claw.

In consequence of this fierce disposition, a native
who has killed one of these animals is held in the
highest esteem by the rest of his tribe, who regard with
envy the necklace of the teeth and claws, and the
'kaross,' or cloak, which he forms from the skin. The

tail, too, is carefully preserved, and dangles from the string which passes round the waist of the successful hunter.

During all its ravages, the animal behaves with a caution which renders it a very difficult matter even to trace the marauder.

He will not approach a farm where he can detect the least sign of the presence of danger, and is even cunning enough to take up his quarters near one village, and commit his depredations in another at a considerable distance, in order to lessen the chance of his retreat being discovered. He often removes to a distant part of the country, too, if he has committed many ravages in his old locality, and fears that he may be in danger in consequence.

Although the size of the leopard is far inferior to that of the lion or tiger, its strength is very great when the dimensions of the animal are taken into account. One of these creatures has even been known to drag a couple of wolf-hounds, which were tethered together, for a considerable distance into the bush, in spite of their struggles. Animals far larger and heavier than itself, too, fall victims to its attacks, and are carried away without apparent difficulty.

The muscular force which is compressed into a leopard's body is really amazing. In his 'Eight Years in Ceylon,' Sir H. Baker has the following remarks on it :—

'The power of the animal is wonderful in proportion to its weight. I have seen a full-grown bullock with its neck broken by a leopard. It is the popular belief the effect is produced by a blow of the paw : this is not the case ; it is not simply the blow, but the combination of the weight, the muscular power, and the momentum of the spring, which render the effects of a leopard's attack so surprising. . . . .

The immense power of muscle is displayed in the concentrated energy of the spring. The leopard flies through the air, settles on the throat, usually throwing his own body over the animal, while his teeth and claws are fixed on the neck. This is the manner in which the spine of an animal is broken, viz., by a sudden twist, and not simply by a blow.'

The same author mentions that he once found a Malabar lad sitting under a tree and looking very weak and ill. He sent some of his men to bring the lad to his house, but when they reached him they found that he was dead. He was buried by the road side, but a few days afterwards it was found that the leopards had discovered the buried body, dug it up, and devoured it. The footprints, which were quite fresh upon the damp soil, afforded unmistakable evidence against the offenders.

Leopards seem to be one of the many hindrances to agriculture in Ceylon.

They are so cunning that it is hardly possible to take effectual precautions against them, and they can hide themselves so easily in the almost impenetrable jungle, that to extirpate them is a hopeless task, unless the whole of the jungle be cleared away. Even then, so great is the power of vegetation, that the neglect of two or three months will permit the jungle to replace itself by fresh growths.

Cattle can hardly be considered safe even when fastened into their houses, for the leopards will clamber on the roof, tear away the thatch, and so gain admission to the shed. Once inside, a leopard will kill every animal in the shed, not for the purpose of satisfying its hunger, but from the mere lust of slaughter.

The cunning of the man-eaters is proverbial. One favourite manœuvre is for the animal to show itself at

F

one end of a village, and make a sham attack upon it. When it has drawn all the armed men in pursuit, it quietly sneaks away, skirts the village under cover, slips in at the other end, pounces upon one of the inhabitants—generally a child—and escapes with its prey into the bush.

The young of the leopard vary from one to five in number. They are pretty little creatures, and as playful as kittens, gambolling with their mother in just the same manner. For the first few weeks of their life the markings are very indistinct, but become more conspicuous as the animals grow older.

Like most of the members of the cat tribe, the leopard has occasionally been tamed, and has sometimes even been allowed to range the house at will, after the manner of a favourite cat. All these animals, however, have been captured when very young, before their savage instincts have had time to show themselves.

This animal is sometimes known as the Panther, the two being merely very slight varieties of the same species.

No VII.—THE CAT TRIBE.

PART III.—THE JAGUAR, PUMA, ETC.

THE Ounce (*Leopardus uncia*) of Asia resembles the Leopard so closely, both in appearance and habits, that a detailed description is unnecessary. It may be distinguished from either of the preceding animals by the woolly aspect of its fur.

We therefore pass it by and come to the Jaguar (*Leopardus onca*) of the American continent. In its native country the animal is usually spoken of as the Tiger, just as the bison is erroneously dubbed the buffalo.

The jaguar is by no means unlike the leopard in form and markings, but may be easily distinguished by one or two peculiarities.

In the first place, two or three bold black stripes are drawn longitudinally across the breast, these being never found in the leopard. The spots, too, with which the body is thickly covered, are more angular in shape than is the case in that animal, and are formed like rosettes, each containing either one or two smaller spots in its centre. Along the spine, from the neck to the first foot or so of the tail, runs a chain of solid black spots and dashes. The tail, too, is much shorter in proportion to the size of the animal, and barely touches the ground when its owner is standing erect.

The ground colour of the jaguar's fur is a bright

tawny brown, in some specimens being much more brilliant than in others. A black variety is occasionally met with, just as is the case in the leopard, the whole fur assuming the dusky hue of the spots, which are very indistinct, and only to be distinguished in certain lights.

Like the tiger, the jaguar is of an extremely cautious nature, and seldom ventures upon an open attack, unless his enemy be very much inferior in size and strength to himself. Should he meet with a herd of animals, or a party of travellers, he will dog their steps for miles, hoping to surprise an individual when separated by a short distance from his comrades.

As regards his food, the jaguar is extremely impartial, and preys alike upon all branches of the animal kingdom. His favourite diet is, perhaps, the flesh of the various monkeys which inhabit the American forests, the attainment of which delicacy, however, is attended with considerable difficulty.

For, though the jaguar is an adept in the art of tree-climbing, and can make his way among the branches with considerable ease and facility, the monkeys are even greater proficients, and, by the quickness and agility of their movements, would soon distance their pursuer if he resorted to open chase. His usual method of obtaining his favourite food, therefore, is by leaping upon the unsuspecting animals from some place of concealment, or by surprising them when asleep. In such a case a few strokes of his powerful paw dash several of the animals to the ground, where their assailant can devour them at his leisure.

Another favourite delicacy is the flesh of the peccary, the procuring of which is fraught with equal difficulty and far more danger than is the case with the monkeys.

For the peccary is distinguished by the possession

of a fierce, unreasoning courage, causing it to dash at the most formidable foe, and wreak its vengeance with its terrible tusks, which cut like so many razors. In fact, if a jaguar were to be attacked by a herd of these little animals, he would have no chance against them, and could only save his life by resorting to a tree until their patience became exhausted, and they retired from the neighbourhood.

The capybara falls a frequent victim to the attacks of the jaguar, who even follows it into the water. Large animals, such as horses or deer, it kills in the same manner as does the leopard, namely, by leaping upon them from the overhanging branches of some tree, and breaking their necks by a powerful wrench with the fore-paws. Even animals of considerable size are carried off without difficulty by the jaguar, which has been known to attack two horses which were tethered together, kill one, and drag both animals to its lair, in spite of the struggles of the survivor.

Of birds, also, the jaguar is fond, and strikes them down with a blow of his paw. Even should his intended prey take to flight, he is often able, by one of his wonderful bounds, to capture it before it has passed beyond his reach. Fish he captures by lying in wait upon the banks of a stream, and hooking them out with his paw as they pass beneath him. Turtles, too, often fall a prey to him, and are killed and eaten in a very ingenious manner.

Watching for the female turtles as they make for the sea after laying their eggs in the sand, the jaguar springs upon them, and quickly turns them upon their backs, a position in which they are perfectly helpless. He then breaks away the softer parts of the shell by the tail, and, inserting his paw, scoops out the whole of the flesh through the aperture thus made. Of the eggs, too, he is very fond, digging them up from the sand in which they were deposited. Lizards, shellfish, and even insects also fall victims to his voracious appetite.

In farmyards the jaguar is a terrible enemy, doing the utmost damage among stock of all kinds. Some of the earlier settlers, in fact, were so troubled by these animals that they found it perfectly impossible to keep any live-stock whatever until the jaguars were finally driven from the neighbourhood. And this was no easy task, the craft and cunning of the animals rendering it a very difficult matter to kill or trap them.

The hunting-dogs of the country show a wonderful aptitude for tracking the jaguar, as is well described by the late Mr. C. B. Brown, in his *Camp and Canoe Life in British Guiana*, as follows :—

'Many of the Indian hunting-dogs, trained for deer or tapir, will hunt tigers (*i.e.*, jaguars). When on the track of either of those animals, should they come across the scent of a tiger, their eager and confident manner of pressing on after the game is immediately changed, and, with the hair on their backs erect, they become cautious and nervous to a degree, jumping at even the snapping of a twig. Abandoning the hunt, they take up the tiger's track, and follow it. But should the huntsman call them from it, or not cheer them on with his voice from time to time, they exhibit great fear, and, keeping close to his heels, cannot be induced to hunt any more in that district for that day.

'On the contrary, if allowed to follow the tiger, they track it up with caution, being fully aware of the cunning dodge practised by that animal; which is, when the dog is close at hand, to spring to one side and lie in ambush until it passes, when with one spring the dog is seized.

'Ordinary dogs would fall a prey to this trap, but not the self-taught tiger-dogs. Their fine powers of scent warn them of their near approach to the quarry, when they advance with great caution, never failing to detect the tiger in time, and when once their eye is upon their enemy it has no chance of escape.

'In its pride of strength the jaguar scorns the dogs, and, with a rush like a ball from a cannon, springs at one of them, feeling sure that it cannot escape.

'It has reckoned, however, without its host, for the dog eludes the spring with ease, and with great quickness flies on the tiger's flank, giving it a severe nip. As the tiger turns with a growl of pain and disappointment, the dog is off to a little distance, yelping lustily, and never remaining still an instant, but darting first on one side and then on the other. After one or two ineffectual charges the tiger gives it up, and on the approach of the hunter springs into the nearest suitable tree, which it seldom leaves alive.'

The jaguar is very tenacious of life, and even when mortally wounded will often travel to a considerable distance before it succumbs to its hurt.

This animal seems more easily tamed than most of the larger members of the Cat tribe, and becomes thoroughly domesticated, following its master like a dog, and allowing all manner of liberties to be taken with it without resenting them.

The Puma *(Leopardus concolor)* is the next of importance among the members of the Cat tribe. This animal is known under a bewildering variety of names, among which may be mentioned the American Lion, the Panther (or 'Painter'), the Cougar, and the Gouazouara. Sometimes, too, it is erroneously termed the Carcajou, which is one of the deer tribe, and also the Kinkajou, a remarkable animal belonging to the ursine group

The title of American Lion evidently refers to the colour of its fur, which is of an uniform tawny hue like that of the animal after which it is named. The tip of the tail is black also, but does not possess the tuft of hair which is a distinguishing characteristic of the true lion.

The puma is by no means so large an animal as any of the preceding, seldom exceeding six feet and a half in total length, of which almost one-third is occupied by the tail. The head is remarkably small, causing the animal to appear even less in size than is actually the case.

It is a rather curious fact that the cubs of the puma should be marked during their infancy with greyish-black stripes, just as are the young of the lion. Besides these, a number of darker spots are visible over the greater part of the body, both stripes and spots disappearing in the course of a few months.

The puma is another of the tree-climbing cats, and its limbs are wonderfully strong and powerful in order to fit it for its semi-arboreal life. Its habits, when in search of prey, strongly remind one of those of the jaguar, crouching, as it does, like that animal, among the branches of some convenient tree until an animal is unfortunate enough to pass beneath. While resting among the foliage it is remarkably inconspicuous, the body being flattened against a bough, and the dark-tawny fur harmonizing almost perfectly with the bark.

Although terribly destructive in farmyards and so on, as many as fifty sheep in one district alone having been known to fall victims to the animal in the course of a single night, the puma is not personally feared by the settlers and hunters. When meditating an attack it is even more cautious than the jaguar, often following its quarry for miles without daring to show itself. Even when it summons sufficient courage to follow up its attack, and arrives within springing distance of its intended victim, it cowers and shrinks away if a bold front be shown, appearing unable to withstand the gaze of the human eye.

The food of the puma is much the same as that of the jaguar, the peccary and the capybara being especial favourites with it.

The various species of Felidæ, called TIGER-CATS, include several pretty and graceful animals, such as the Ocelots, the Margay, the Rimau-Dahan, and the Chati.

The Ocelots are all inhabitants of tropical America, the most abundant species being that known as the Common Ocelot (*Leopardus pardalis*). This animal is a singularly pretty one, some four feet in length, from the nose to the tip of the tail, and standing about eighteen inches in height at the shoulder. The ground colour is a delicate greyish fawn, marked with broken bands of darker fawn edged with black. Along the spine runs an unbroken black line. The ears are black, with the exception of a white spot upon the back.

In consequence of the handsome markings and delicate fur, the skins of these animals are much sought after.

Another species is the Grey Ocelot (*Leopardus griseus*), which is of a lighter colour than the preceding animal, the spots, also, being less numerous and distinct. All the species are very quick and active in their movements, and in form and habits strongly resemble miniature leopards.

Our own domestic cat, really a descendant from the Egyptian animal (*Felis maniculata*), is generally supposed to have sprung from the Wild Cat (*Felis catus*), which at one time was very abundant in this country. In the olden times, when hunting and warfare were almost the sole occupation of the upper classes, this animal was even preserved for the chase, just as is the fox at the present time, and severe penalties were enacted against those who should cause its destruction, except in the legitimate manner.

At the present day, the Wild Cat is almost extinct, as far as regards Great Britain, a few scattered specimens, only, existing in some of the Scottish mountain woods. There it causes considerable havoc amongst the game, just as our domestic cat will if it once imbibes a taste for poaching. In fact, there is hardly a more inveterate enemy to partridges and pheasants than a pet cat, which will visit the coverts night after night, and destroy the birds in numbers, often paying the penalty with her own life should the keeper happen to meet with her. Many tame cats, even, leave their homes, and take entirely to a wild life, living on the game which they capture. When trapped, these are often mistaken by the keepers for the genuine wild species. It is hardly possible to pass through a preserve without noticing the dead bodies of several cats, which have been shot by the gamekeepers, and hung up on the 'keepers' trees' in company with the carcases of weasels, stoats, etc., as a warning to other 'vermin.'

The differences between the wild and the domestic cats are very apparent. The markings of the former vary but very little, the ground colour being an uniform yellowish grey, while a number of dark streaks run round the body at right angles to the line of the body and limbs, reminding one rather strongly of the markings of the tiger. A chain of black spots runs down the spine as far as the tail, which is very much shorter and more bushy in proportion than that of the domestic species. The tip, for an inch or so, is invariably black, the rest being banded in the same way as the body. The ears, too, are much shorter than those of the tame animal, and the aspect of the creature is remarkably fierce and savage. Its character is not belied by its appearance, for its courage and ferocity are so great that even to an armed man it is no contemptible foe.

The Lynxes form a very conspicuous group. The Common Lynx *(Lyncus virgatus)* of Southern Europe is well known by name, its keenness of sight having passed into a proverb. It is spread over a considerable part of the mountainous parts of South Europe, and is also found in many of the forests of Northern Asia.

The Lynx is not a very large animal, being about three feet in total length, exclusive of the tail. The fur is of a dark-grey colour, varying slightly, according to the season of the year, with darker spots of various sizes. The tail is very short, sometimes being barely six inches in length. The ears are remarkable for the tuft of long hair which fringes the tips.

The Canadian Lynx, or Peeshoo *(Lyncus Canadensis)* is chiefly noticeable for its mode of running, which operation consists of a series of bounds, all four feet coming to the ground almost simultaneously. The fur, which is long and fine, is much sought after.

There is another species, the Booted Lynx *(Lyncus caligatus)* found both in Asia and Africa, which is so called from the deep black colour of the lower parts of the legs, causing them to appear as though enclosed in tightly-fitting boots. None of the Lynxes are particularly destructive, except to the small animals, such as hares, rabbits, etc., upon which they prey.

One of the most interesting of the Cat tribe, and the last which can be mentioned in this paper, is the well-known Chetah, or Hunting Leopard *(Gueparda jubata)*, sometimes termed the Youze.

The Chetah is found both in Asia and Africa, but in the former continent alone has been brought under the dominion of man. In size it is slightly superior to the leopard, its long limbs causing it to appear even larger than it is in reality. The markings are not unlike those of the leopard, to which animal, however, it is by no means closely related, appearing to form a connecting link between the feline and canine races. It is only slightly possessed of the power of climbing trees, and its limbs do not possess the strength of most of the cats.

Unlike most of its tribe, the Chetah captures its prey by open chase, and it is in this way that it is made useful to man, the animal being carefully trained for capturing game by many of the Asiatic races.

For this purpose, the animal is blindfolded, and taken out to the scene of operations in a light cart, where he is kept until a herd of deer or other game comes within sight. The hood is then removed, and the animal's attention directed towards the quarry. The Chetah immediately slips gently off the car, always doing so on the side which is away from the deer. Flattening his body on the ground, he creeps up to within a short distance of the unsuspecting animal, taking advantage of any bush or stone as a cover. He then launches himself upon the doomed animal, seldom needing more than two or three springs, fastens on its neck, and pulls it to the ground. The keepers immediately hurry to the spot, and take off his attention by offering some dainty, such as a ladleful of the blood. The slaughtered animal is then secured, and the Chetah is hooded and led back to the car, when he waits for another victim.

The Chetah is very easily tamed, being naturally of a very gentle and placid disposition. Even a newly-caught individual is easily managed, seldom or never exhibiting the savage nature of the lions and tigers, and other members of the Cat tribe.

No. VIII.—THE DOG TRIBE.

U PON examining the respective skeletons of one of
the cats and of a member of the dog family, we
shall see that the distinctions are very apparent. The
form of the cat is evidently that of an animal which is
intended to creep up stealthily to its intended victim,
and then to despatch it by means of a single bound.
Almost every detail of the cat tells the same story ; the
structure of the skeleton, the pads beneath the feet,
the retractile claws, and the powerful fore-limbs all
unmistakably pointing to the same conclusion.

But the members of the dog tribe are as evidently
intended for the pursuit of prey by rapidity of foot
alone. The limbs are formed more for speed than for
strength, and the head and shoulders do not possess the
massive power found in the cats, which bury their
teeth in the flesh of the victim, and retain their hold
until the death-blow is given.

The members of the dog tribe, including the wolves,
jackals, and foxes, differ from the cats in many points
both of structure and habits.

It will be remembered that the latter animals feed
chiefly upon prey captured by themselves ; the dogs, on
the contrary, will devour with equal avidity the flesh
of any slaughtered animal which they may find, as well
as offal of all sorts.

A glance at the feet will at once point out the

reason. The claws, which in the cats are of such service in the capture of prey, are in the dogs of comparatively little use for this purpose, being short, blunt, and non-retractile.

That there must be some original type of the dog is self-evident, though it is almost, if not quite impossible to ascertain with any precision what that type may be.

There is no similar difficulty about the wolves, foxes, and other members of the dog tribe, because they live a wild life, and so can preserve their own typical character. But the very nature of the dog compels it to withdraw itself from a wild life, and attach itself to man. More or less it becomes his companion, and does his work, and it is therefore necessarily modified according to the race, the climate, and the domestic economy of the human beings with whom it associates itself.

Take the cities of the East.

There we have the dog, not quite, but nearly wild, doing the work of man, by acting as scavenger, and so enabling man to live.

Take the semi-nomad North American Indian, who lives in tents, but remains for months, and sometimes for years, in the same locality. He is a warrior and hunter, and nothing else, utterly despising work, and, even if he does grow a crop of maize, delegating all the work to the women.

With him the dog becomes the guardian of the temporary village. He knows every inhabitant, and allows no stranger to enter unless accompanied by one of the warriors.

Go further north, and take the Esquimaux, an aggregation of equally nomad tribes, but inhabiting a region of almost perpetual snow and ice. Here the dog becomes the beast of burden and traction. He can have no roof to cover him, for his masters are themselves glad to huddle in their little huts of snow, which are hardly large enough to hold them and their families. So he is supplied with a coat of long and dense fur, which enables him to live where the smooth-haired dog would be frozen to death.

Should he belong to a pastoral race, he becomes the faithful guardian of the flocks. Should he belong to a race that lives by hunting, and has to contend for food with the wild beasts, he becomes the fierce and tireless hound. And should he belong to those who only want him to make a pet of him, he becomes a pet accordingly, useless, silly, and selfish.

All the varieties of the domestic dog are purely artificial, and as in these pages we treat of Nature, and not of art, we make no mention of them.

These, therefore, will be altogether omitted, and the wild animals of the tribe alone taken into consideration.

The first to be mentioned is the well-known Dhole, or Kholsun (*Cuon Dukhuensis*), which is found in the more western parts of British India.

The colour of the animal is a dark bay, the muzzle, ears, and tip of the tail being darker than the rest of the body. In size, it about equals a rather small greyhound.

Common though the dhole is in the country which it inhabits, it is seldom or never seen by the residents, owing to its timid and retired mode of life. By many travellers, indeed, it has been considered as merely a myth of the natives. In the dense jungles, however, it is abundant enough.

The most noticeable point concerning the dhole is its fondness for the chase. For the purpose of procuring prey it combines in large packs of some fifty or sixty individuals, and by sheer force of numbers contrives to overcome such large and powerful game as the wild boar and the tiger. And this is the more remarkable when we consider what insignificant weapons the dhole can bring to bear against the powerful tusks and talons of its adversaries. The secret of success, however, lies in its courage and pertinacity, for, although their comrades are being struck down on all sides, the survivors continue the attack without allowing their foe an instant's rest, and do not cease their onslaught until the unfortunate animal yields from fatigue and loss of blood.

The speed of the dhole is very considerable, even the swift-footed deer being unable to escape from their apparently insignificant pursuers. It is a curious fact that, while engaged in the chase, the dhole is almost silent, an occasional low whimper being the only sound ever emitted.

In Nepaul, and the northern parts of India, an animal nearly allied to the dhole is found, which is generally known as the Buansuah (*Cuon primævus*).

This animal, which is generally supposed to be the progenitor of our domestic dog, is very similar in habits and general appearance to the animal already described. Like the dhole, it hunts in packs, which, however, seldom consist of more than ten or a dozen individuals. It differs also in its habit of giving tongue while running, continually uttering a peculiar bark, very distinct from that of the domesticated animal.

The buansuah is often captured when young, and carefully trained for the chase, the wild boar being the selected quarry. For the purpose of hunting this animal the buansuah is very valuable, its sudden, snapping bite being far more effective than the attack of the ordinary hound. It is not so easily taught, however, to follow other game, being rather apt to relinquish the pursuit almost at the moment of capture.

Jackal

G

The well-known DINGO (*Canis dingo*) of Australia must not be passed by without mention.

This animal is not thought to be an indigenous inhabitant of the continent it inhabits, inasmuch as all Australian mammals seem to be marsupials, but is supposed to have been imported from some unknown source many years ago. It is rather a handsome animal, being of a rich reddish-brown colour, sprinkled with blackish hairs over the greater part of the body ; the ears are short and erect, and the tail is thick and bushy, almost as much so as the well-known ' brush' of the fox.

To the colonists and farmers of Australia the dingo is an unmitigated pest, ravaging the flocks night after night, and committing incalculable damage in a very short space of time. As many as twelve hundred sheep and lambs have been stolen from a single colony by these animals in the course of three months. And the cunning of the dingo, being little inferior to that of the fox, renders it a very difficult matter for the settlers to protect their herds from the attacks of the wily foe.

Like the dhole of India, the dingo hunts in large packs, each of which has its appointed sphere of action, and never trespasses into the district of another band. When attacked by human foes, it shows little inclination to fight except when brought to bay, when it will attack its pursuers with great ferocity.

Various attempts have been made to domesticate the dingo, and with partial success ; but its temper is always very uncertain, and it is always apt to attack any passing human being, its own master not excepted, without the slightest provocation or apparent cause.

LEAVING the dogs themselves, we come to the closely allied JACKALS, which are found in many parts of the African and Asiatic continents. There are

several species of these animals, of which the most
abundant and familiar are the common Jackal, or
Kholah (*Canis aureus*) of India and Ceylon, etc., and the
Black-backed Jackal (*Canis mesomelas*) of Southern
Africa.

The former of these animals is found in very great
numbers, but, owing to its nocturnal habits, is more
often heard than seen, keeping up, as it does, a perpetual howl from dusk to dawn. It usually herds together in packs, which retire to the thick forest during
the day, and sally out after dark in search of food.

Their prey usually consists of the smaller quadru
peds, which they can overpower without much diffi
culty. They are not very particular, however, and are
equally satisfied with the carcase of any slain animal
which they may happen to fiud.

The jackal is often known as the ' lion's provider,' owing
to its habit of following closely upon the footsteps of the
large members of the feline tribe. This title, however,
is rather misapplied, for jackals follow a lion or tiger
solely for the purpose of preying upon what remains of
the carcases of his prey after his lordly appetite is
satisfied. A ring of jackals may often be seen surrounding a lion when engaged in feeding, patiently
waiting until his wants are supplied and they can con-
sume the remainder.

Occasionally a jackal will separate himself from his
companions, and live in solitude. These hermit
animals are terrible foes to the farmers, attacking the
hen-roosts and sheep-folds by night, and causing great
havoc amongst the assembled animals.

The fur of the jackal is of a yellowish-brown tinge,
whence the scientific name, *aureus*—*i.e.*, ' golden '—is
derived. In size, it rather exceeds the common British
fox. Like that animal, it is possessed of a powerful
and unpleasant odour, which, singularly enough,

Dingo.

gradually dies away if the creature be kept in confinement.

The black-backed jackal of South Africa may be easily distinguished from the Asiatic species by the black and white markings upon the back. The size and general appearance of both animals are much the same; in habits also they are so similar that a detailed description is rendered unnecessary.

THE fiercest and most terrible animals of the dog tribe are found in the WOLVES, which inhabit almost all parts of the world, from the Arctic regions to the tropics.

There are several kinds of wolves, as well as many varieties, which by some authors are elevated to the rank of species. The best known of these is the Common Wolf (*Canis lupus*), which is so abundant in many parts of Europe. The colour of this animal is grey, rather thickly sprinkled with black hairs, and tinted in some parts of the body with a warm fawn hue; the lower parts of the body are almost white.

When found singly, which is not very often, the wolf is a comparatively insignificant enemy, his courage not being of a very high order; when banded together in packs, however, which is almost always the case, there are few animals which he cannot overcome. Even the bear himself often falls a victim to his attacks, and such powerful animals as the buffalo and the elk have little or no chance against him.

One great peculiarity in the wolf lies in its unwearying pertinacity when engaged in the pursuit of prey. Once fairly upon the trail, it follows up the victim with a long, swinging gallop, which carries it along at a wonderful pace, and is certain, sooner or later, to bring it up with the quarry, however fleet the hunted animal may be.

When the victim is once overtaken, its chance of escape is small indeed. The wolves crowd round it

Wolf.

attacking it with a series of fierce, snapping bites, each of which causes the teeth to meet in the flesh of their adversary. If one animal is killed, another at once takes its place, and before very long the issue of the struggle is decided.

When the victim is once slain, the wolves seem to lose all control over themselves, fighting fiercely for every morsel of the coveted flesh, and attacking each other with the most ungovernable fury. If one should be overcome, he is instantly devoured by the survivors, and it is even reported that any animal who is unfortunate enough to dabble himself with the blood of the victim is certain to share the same fate. A weak and sickly wolf, also, is sure to fall a prey to the ravenous hunger of its comrades.

The wolf is not very particular as to the nature of his prey, animals of all kinds, even to frogs, toads, and insects, supplying him with food.

It seems strange that so bold an animal as is the wolf as a general rule, should at other times exhibit the most utter cowardice. If a wolf is caught in a trap, for instance, its courage seems at once to leave it, and it cowers down in a corner of its prison, and allows itself to be slaughtered without offering the slightest resistance.

As is the case with the lion, too, its suspicious nature sometimes offers a chance of escape to its intended victims. Travellers, when chased by wolves, have more than once escaped by trailing a piece of rope, or some other object from the carriage, and changing it for another as soon as the wolves began to lose their suspicions.

The hunters also take advantage of this excessive caution, and protect their slaughtered game from the wolf as they do from the lion—viz., by planting a stick by the side of the carcase, and attaching to it a streamer of white cloth, which flutters in the wind, and deters the fierce animals from approaching.

The Black Wolf (*Canis occidentalis*) of America greatly resembles the last-mentioned animal, both in character and habits. In appearance also it differs only in a slight degree, and for a long time was considered to be nothing more than a permanent variety of the common species.

A smaller and more abundant animal, found in great numbers upon the vast American plains, is the Prairie Wolf (*Canis latrans*). These animals are always to be seen in great profusion upon the outskirt of the herds of bisons which populate the plains, hovering in the neighbourhood in the hopes of overcoming any injured or weakly member of the herd. A considerable number also usually follow the hunter, feeding upon the carcases of animals which he has slain, and from which he has taken sufficient for his own requirements.

One of the best-known of the American wolves is the Coyote, or Cajote (*Canis ochropus*), which is equally hated and despised by the hunters on account of its skulking and cowardly nature. This animal, which is very abundant on the prairies, has more of a fox-like aspect than the other wolves. In general habits it presents no very great difference from the previously-mentioned species.

The young of the wolves vary from three to eight or nine in number, and are brought up in a kind of nest constructed by the mother, which is lined with moss and fur pulled from her own body. When they attain the age of six or seven months, the young wolves are able to take care of themselves.

In spite of their fierce and savage nature, wolves have occasionally been tamed and brought into subjection—such animals, of course, being captured when quite young, before their character was fully developed. A mixed breed has sometimes occurred, between the tame wolf and the domestic dog, their offspring being especially powerful and courageous.

NEXT we come to the FOXES, of which there are several species. Formerly included by zoologists in the preceding genus *Canis*, together with the dogs and the wolves, they were separated by later writers on account of the elongated pupil of the eye, and also from the bushy nature of the tail. The ears, too, are always triangular, and are sharply pointed.

The best known of the foxes, of course, is that found in our own country (*Vulpes vulgaris*), and which is so familiar to us on account of the chase, for which it is specially preserved.

The colour of this animal is a rich reddish-brown, becoming rather lighter on the lower parts of the body. At the approach of winter the fur becomes perceptibly paler, and at the same time increases greatly in thickness, just as is the case with the stoat, although not to the same degree. The tip of the tail, or 'brush,' always retains a more or less whitish hue.

Perhaps the most remarkable point in the nature of the fox is the singularly powerful and unpleasant odour which is exuded from the body, and which proceeds from glands situated near the tail. So strong is this scent, that any object touched by the fox retains the odour for a considerable period of time.

The fox seems to be aware of the possession of this peculiar property, although, in all probability, his nostrils are unable to perceive the odour ; for when hunted he will try every means which occurs to his fertile brain to break the line of scent. For this purpose he employs a perfect variety of tricks, such as returning upon his own track for some little distance,

The Fox.

and leaping off at right angles, in the hope of escaping before the fraud is discovered. The animal will even roll in any odorous substance he can find, in order to disguise his own peculiar scent, and mislead the hounds by causing them to imagine that they are upon the wrong track.

Many foxes become so crafty that they make their escape again and again, always contriving to elude the pack, until the hounds become completely dispirited, and consider the issue of the chase as a foregone conclusion.

There is a gravel pit in Kent which exhibits the cunning of the fox in a very singular manner.

The animal has burrowed into the ground at some distance from the mouth of the pit, carefully concealing the entrance to the ' earth ' among the tangled vegetation. Carrying the tunnel on, a second exit appears in the side of the pit itself, some half-way to the ground.

When hunted, the animal was evidently accustomed to enter his burrow at the upper end, pass through it, and make his escape by leaping into the quarry, while the hounds were at fault above, the idea of the second exit not being likely to strike the huntsmen, at any rate for some little time.

The same craft and cunning is employed by the fox when pillaging the hen-roosts, etc., of the neighbourhood, his visits being paid with such caution that detection is rendered almost impossible.

Yet, cunning as is the fox as a general rule, on some occasions his craft seems almost entirely to desert him. The late Mr. Charles Waterton, in one of his well known essays, relates an instance of this want of sagacity.

A fox, visiting a poultry-yard, had made off with eight young turkeys. Finding that his booty would more than suffice for a single meal, he buried five of his victims in a neighbouring garden, evidently intending to return on the following evening and resume his banquet. But although the bodies of the slaughtered birds were carefully concealed, one wing of each was left projecting above the soil, thus pointing out the transaction to every passer-by. As Mr. Waterton remarks, ' An ass, in this case, would have shown just as much talent and cunning as Reynard himself had exhibited.'

Passing to the foxes of other countries, the American fox (*Vulpes fulvus*) deserves a passing mention. This animal is very variable in its colouring, specimens having been found of almost every intermediate hue between black and pale yellow. A black streak almost invariably crosses the shoulders, earning for the animal the alternative title of ' cross fox.'

The Arctic fox (*Vulpes lagopus*) is a very well-known animal, chiefly on account of the valuable fur, which is much used in commerce. During the winter, at which time it is most in request, the coat is of a beautiful silky white, darkening to a dull greyish-brown as the season advances. The Arctic fox inhabits the northern regions of Europe, Asia, and America.

This animal appears to be almost destitute of the remarkable cunning of the others of its race, being easily trapped, and allowing a hunter to approach within easy shooting distance. In one way, however, it is sagacious enough, possessing the power of imitating the cries of the birds upon which it feeds, and so enticing them within its reach.

SOMEWHAT resembling a very small fox in general appearance, the Asse, or Caama (*Vulpes caama*) merits a passing mention.

This animal is found in Southern Africa, where it is remarkable for its inroads upon the nests of the ostriches the eggs of which it destroys in great numbers. Not being able to pierce the thick shell with its tiny jaws, it rolls the egg against a stone, or other hard substance, and so contrives to obtain the contents.

Passing by the Otocyon and the Fennec, we come to the last of the dog tribe which can be mentioned in this paper, and whose position in the family is as yet very uncertain. This is the Hunting Dog (*Lycaon venaticus*), which has been thought to constitute a connecting link between the dog tribe and the hyænas ; a final decision, however, has not as yet been arrived at. In fact, the characteristics of the hyænas and the dogs are so curiously intermixed in this strange animal, that it must be a matter of extreme difficulty to relegate it to its true position in the scale of creation.

The colour of the hunting dog is a reddish-brown, mottled with black-and-white patches ; the nose and jaws are black, and a black streak runs along the head between the eyes. The ears are large, and the tail is long and bushy.

Like the dhole and the buansuah, the hunting dog combines in large packs for the purpose of procuring game, generally choosing the night-time for its predatory excursions. Its sense of scent is wonderfully keen, and its speed very great, and it is but seldom that the hunted animal is allowed to escape.

It will be seen that, although the cat and dog tribes both include some of the larger carnivora, the two families are, in structure as well as in habits, essentially different ; and that the distinctions between the domestic cat and dog are no greater than between their more savage relatives which have never known the loss of freedom.

No. IX.—THE CETACEA, OR WHALES.

THE mere fact that the Whales are exclusively inhabitants of the water, is usually held to be a sufficient proof that they should be included among the fishes. In fact, it is generally considered that all sub-aquatic creatures, be they mammals, fish, crustacea, or radiates, may be included under the one comprehensive title.

As far as the whales are concerned, there is certainly some ground for the idea. Their habits and mode of life, their food, and their very form, so closely resemble those of the fishes that we can scarcely wonder if these animals are popularly supposed to form part of that group. We have only to look a little more closely into their structure, however, to find that they have nothing in common with fishes.

An examination of their mode of breathing is alone sufficient to point out the true position of the whales in the animal kingdom.

It is well known that all the fishes respire by means of gills, by the agency of which the necessary oxygen is extracted from the water which they inhabit. But the whales, like all other mammals, are obliged to breathe atmospheric air by means of lungs, for which purpose they are compelled to rise to the surface of the water. Were they prevented from doing so, they would be drowned just as would any other mammal under similar circumstances. The mode of life of the whales, however, differs completely from that of other mammalia, the breathing apparatus being modified in such a manner as to allow them to remain beneath the water for a considerable space of time. This structure we shall presently examine in greater detail.

As the method of respiration effectually disproves the general notion that whales should be ranked among the fishes, there is little difficulty in placing them in their true position. The structure of the heart, which possesses two auricles and two ventricles, whereas the fishes only possess one of each, and the fact that the young are nurtured by the mother's milk, are proofs amply sufficient to determine their true situation to be among the mammalia.

In few animals do we find the structure more curiously modified to suit the conditions of existence than is the case with the whales. Passing the whole of their life in the water, their form, like that of the fish, is that best adapted for passing through their native element ; the organs of locomotion, however, are of a different nature.

The fore-limbs, until stripped of their covering, closely resemble the fins of a fish. They are, however, of little use in forcing the animal through the water, their chief duty lying in preserving the equilibrium of the body and in clasping their young. The hinder limbs are not developed, being visible merely as small and imperfect bones when the skeleton is examined. In fact, they can scarcely be said to exist at all, and the very pelvis is only a slight rudimentary process, not attached to the spine.

The great organ of locomotion is found in the tail, which is set transversely with the body, and is usually of very great comparative size. In a whale of ordinary dimensions, the tail, though only a few feet long, would measure in breadth almost one-fourth of the whole length of the body. Even these dimensions, large as they are, are sometimes exceeded; a whale captured a year or two since, and which was only sixty-five feet in length, measured no less than twenty-seven feet in the breadth of the tail.

The muscular power of this organ is simply enormous, the animal being enabled by its aid to leap clear out of the water to a height of several feet, a movement usually known as 'breaching.' The chief danger in whaling lies in the blows of the animal's tail, any one of which is sufficient to dash the boat and its occupants to fragments.

Although in almost every pictorial representation of the whale the eye forms a very conspicuous object, it is in reality extremely small in comparison with the size of the animal, sight being of little use in taking the prey. The ear, too, is exceedingly minute, and for a very good reason.

It will be remembered, of course, that water, which is much more dense than air, is a proportionately excellent conductor of sound. If a man submerges his head he can hear the beat of oars upon the surface, while the boat to which they belong is a mile or more distant.

Again, if a swimmer dives beneath the water, and a heavy blow be struck upon its surface above him, he not only hears the sound, but is almost stunned by the shock. So we can easily see that if the ear of the whale were proportioned in size to the dimensions of its owner, the animal would inevitably be killed by the shock caused from the first blow of its own tail.

In point of fact, the external ear of the whale is so small that it will scarcely admit a crowquill.

Here we are met by another problem.

Even when the whale lies on the surface the ear

Greenland Whale (*Balaena mysticetus*).

is under water, and can only hear sounds that are transmitted through the water. How, then, is the whale to hear sounds that are made above the surface and are transmitted through the air? The difficulty seems insuperable, but is overcome in the simplest manner imaginable. Let us see what is the structure of the ear in mammalia, taking our own as an example.

First, there is an aperture for the admission of air. At a variable depth in this aperture, a very elastic membrane, called the 'tympanum,' or drum, is stretched tightly across it, and is acted upon by any vibrations of air which are rapid and regular enough to become sounds.

On the other side of the drum is a set of bones, called, from their appearance and office, the hammer, anvil, and stirrup. These take the vibrations of the drum, and transmit them to the nerves of hearing, through which they pass to the brain. I may incidentally mention that the modern telephone is nothing but a rude imitation of the structure of the ear.

The tube does not end at the drum, but passes on, though very much reduced in size, to the back of the throat. If this secondary tube (called the Eustachian tube) be stopped, deafness results, because the vibrations of the drum are checked.

Now, in the whale the size of the two tubes is reversed. The external tube is very small, but the Eustachian tube is very large, and passes into the nostrils, or 'blow-hole.' The aperture of this blow-hole is always above water when the whale floats on the surface, so that the vibrations of the air can pass through it to the tympanum. Thus the whale hears through the blow-hole any sounds which are caused by the vibration of air, and through the external tube those sounds which are caused by the vibration of water.

Spermaceti Whale (*Catodon macrocephalus*).

I mentioned that the blow-hole has only a partial right to the name of nostril. It performs only one duty of a nostril, *i.e.*, that of admitting air to the lungs, and is not in any way an organ of scent. In fact, the sense of smell is absolutely wanting in the whale tribe, the entire system of olfactory nerves being absent.

No water can pass down the blow-hole, a simple and very effective valve being so arranged that it closes the aperture by the mere pressure of the water above it.

The whales being warm-blooded animals, some provision must necessarily be made for retaining the vital heat of their bodies in the conditions under which their lives are passed. Yet to all outward appearance, this seems to have been entirely neglected, the smooth and polished skin being apparently the very worst medium which could possibly have been chosen.

A glance beneath the surface, however, tells a different tale. We find that immediately beneath the skin is a layer of coils of fat, some twelve to eighteen or more inches in thickness, which is enclosed in tough, membranous cells. This substance, commonly known as 'blubber,' serves a double purpose, the non-conducting fat retaining the heat of the body, while the thick elastic mass resists the enormous pressure of the water at the vast depths to which the animal descends.

We now come to the remarkable modification of the breathing apparatus which allows the whale to remain beneath the water for a considerable space of time without rising to the surface in order to obtain a fresh supply of air.

As is the case with all warm-blooded animals, respiration in some form or other must be continually kept up. The blood must be constantly supplied with oxygen, or life cannot be preserved.

With the whale, however, the necessity for constant respiration would entirely prevent it from pursuing its search for prey at the depths to which it descends, and would oblige it to face death in one of two forms —starvation or suffocation.

A most wonderful structure is therefore provided, which enables the animal to aërate a supplementary stock of blood, which can be introduced into the circulatory system as occasion requires, taking the place of the exhausted fluid, and doing away with the necessity for constant respiration. This is managed as follows :—

When the whale ascends to the surface of the water in order to breathe, it makes a succession of inhalations, generally some forty or fifty in number, which are usually termed the spoutings, on account of the shower of water mixed with hot breath, which is thrown up into the air to the height of eighteen or twenty feet. During this operation the whole of the blood is thoroughly aërated, not only that in the circulatory system, but also the reserve supply, which is stored away in a vast mass of auxiliary blood-vessels which line the interior of the chest. These vessels contain a sufficient stock of the purified blood to sustain the animal for a considerable time without obtaining a fresh supply of air, and it is by no means unusual for a whale to disappear beneath the water for upwards of half-an-hour without rising to replenish its stock.

Were it not for the knowledge of this habit the difficulties of whaling would be greatly increased; as it is, however, the huge animal is slain with comparatively little trouble.

When a whale is seen, a boat puts off and makes for the spot as speedily as possible.

As soon as the boat approaches within a short distance, a harpoon—a spear with a barbed head, to the end of which is attached a coil of rope—is flung at the animal.

The frightened whale instantly dives beneath the surface, carrying the harpoon with it, the rope being uncoiled from the boat as rapidly as possible. For half an hour or so, the animal remains beneath the surface, but is at length obliged to rise in order to procure a fresh supply of air. No sooner does it appear, and begin its spouting, than the boat approaches, and again drives it below before the operation is completed.

Not having been able to aërate the whole of the blood, it cannot remain so long beneath the surface, and is soon obliged to again rise in search of air. Again it is driven below, and so on until the animal is so weakened from want of air that the pursuers can come to close quarters.

The depth to which a whale will descend when pursued is simply astonishing. On one occasion the animal took down with it more than one thousand fathoms of rope, or considerably over a mile, and yet was enabled to bear the tremendous weight of the rope, and also to drag the boat with its pursuers rapidly through the water.

It appears strange that so comparatively insignificant a weapon as the harpoon should prove so deadly to an animal of such enormous dimensions. But the animal does not lose its life on account of the very slight wound produced by the harpoon.

The real instrument of death is the spear, which has a small and very sharp blade and a very long handle. When the whale is quite exhausted by fatigue, the spear is thrust into the vital organs, and in spite of its size the animal easily succumbs.

Formerly, the harpoon was always thrown by hand. It is now mostly shot from a gun, and, of course, can penetrate more deeply than the hand-thrown weapon.

To mankind, whether in a civilized or a savage condition, the whale is of inestimable value. From the blubber and other parts of the body we obtain the valuable oil, which in many countries is almost a necessity of life ; the so-called 'whalebone,' and the bones themselves are of considerable value ; and, by the dwellers of the polar regions, almost every part of the body is used as food, the skin, and, more particularly, certain parts of the gums, being considered as very great dainties.

The curious substance popularly known as 'whale-bone' deserves a few words.

To the title of 'bone' it has no claims whatever, its structure being analogous to that of hair, feathers, scales, and teeth, which are merely the same substance under different forms. It is found in the jaws, lying in thin flat plates of various breadth, and from ten to twelve feet in length. These do not spring from the gums themselves, but from a curious vascular formation resting upon them. Each plate is split at the extremity into a number of hair-like filaments.

In a certain sense, the whalebone takes the place of the teeth, inasmuch as it captures the prey, although it is not used for mastication, which, from the nature of the food, is rendered unnecessary. Its use is as follows.

Those species of whales which are provided with the 'baleen,' or whalebone, prey upon creatures of very minute size, such as small shrimps, crabs, and lobsters, medusæ, etc., which are generally found in large shoals. Its chief food consists of a small mollusc called the Clio. Opening its huge jaws to the widest extent, the whale drives rapidly through the shoal, thus filling the mouth with the little creatures ; the jaws are then closed, and the contained water is driven out through the interstices of the whalebone. This, how-ever, completely prevents the escape of the prey, which can then be swallowed at leisure.

Having now glanced at the principal characteristics of the whales as a family, we will take each of the more important members in turn, and devote a short space to their habits and peculiarities.

The first in order is the Greenland Whale *(Balæna mysticetus)*, or, as it is often termed, the Right Whale. This whale is an inhabitant of the seas bordering upon the northern polar regions, where, in spite of the annual slaughter, it is still to be found in considerable numbers.

The Greenland whale, although of great size, is by no means equal to the huge rorqual in its dimensions. Its average length is from fifty-five to sixty feet. The head is extremely large, occupying rather more than one-third of the whole bulk.

The colour of this whale · is a deep velvety black upon the upper parts of the body, and greyish white upon the under surface. It is one of the most useful of all the whales to mankind, the baleen, or whalebone, being long and of fine quality, and the oil rich, and found in great quantity. Even the very bones teem with the oil, the jawbones especially producing a considerable amount.

It is believed that one cub only is produced at a birth in the case of the Greenland whale. For the first few month of its life the baleen is not developed, and the young whale is obliged to depend for the whole of its nourishment upon its mother, who never leaves it until it is old enough to forage for itself.

The Rorqual *(Physalus boops)* is the largest of the whale tribe, sometimes attaining to the extraordinary length of one hundred feet, or even more. In spite of its huge size, it is of comparatively little value, the oil obtained from the body being very scanty, and the whalebone short and of very inferior quality. The animal is therefore seldom molested except by inexperienced sailors who are unable to distinguish it from the Greenland species.

The food of the rorqual consists not only of the minute creatures before mentioned, but also of the larger fish, such as the cod, etc. The nature of the food being so different, the gullet of this whale is of much greater size than in the Greenland whale, in which animal it barely exceeds two inches in diameter. There is a popular saying among seamen that the Greenland whale can swim a jolly-boat and crew in its mouth, and yet be choked with a herring.

In search of its prey, the rorqual often follows the shoals of fish from place to place, and occasionally takes up its quarters upon the borders of the fisheries, to which it causes considerable damage. In such a case it often happens that, pursuing its wished-for prey rather too rashly, it becomes stranded upon the beach, where it is utterly helpless, and is easily slain. A year seldom passes, even in our own country, without a rorqual or two being stranded upon the shores.

The rorqual may be distinguished from the Green-land whale by its dark-greyish hue, by its more slender form, and by the fact of its possessing a dorsal fin. The skin lies in deep longitudinal folds along the under parts of the body, for which reason the name ' rorqual ' was given to it, that title being derived from a Norwegian word signifying a ' whale with folds.'

The Spermaceti Whale *(Catodon macrocephalus)*, or Cachalot, is of great value to mankind, both on account of the oil procured from the blubber, which is of a very fine quality, and also of the substance known as spermaceti, which is found in considerable quantities.

This whale differs in several important points from the two preceding species.

The head is extremely large, occupying nearly a third of the entire length, whence the name *macro-cephalus—i.e.*, ' large-headed '—is derived. The snout is abruptly squared off, and the blow-hole is placed upon the fore part of the head. The jaws are not provided with the baleen, or whalebone, but are furnished instead with a number of formidable teeth, set in the lower jaw, and fitting into corresponding cavities in the upper one. The upper jaw has merely a short row upon each side.

Although to us these teeth are of no particular value, they are held in the greatest esteem by certain savage tribes. On more than one occasion a war has been waged by one chief upon another, merely for the possession of a single whale's tooth.

The cachalot attains to a considerable size, its average length being from seventy to seventy-five feet in length.

The skull, which is elongated and narrow, does not occupy more than one-half of the space assigned to the head, the upper portion being composed of tendinous cells. In two great cavities in this mass is contained the spermaceti, which is found in a fluid, oily condition, and is literally baled out by means of buckets, a hole being cut in the upper part of the head, and the spermaceti extracted just as is water from a well.

The oil expressed from the blubber is of a very fine quality, and is obtained in considerable quantities, a cachalot of ordinary size yielding about one hundred barrels, as well as twenty-four barrels, or thereabouts, of the spermaceti.

This whale is able to remain beneath the water for a much longer period than the previously described species, an hour sometimes elapsing before it is obliged to return to the surface The ' spoutings ' are from sixty to seventy in number, and occupy about ten minutes. It is a curious fact that the number of spoutings is always exactly the same in the same individual.

As regards the locality in which it is found, the cachalot is a rather ubiquitous creature, inhabiting all parts of the ocean, excepting those in the neighbourhood of the polar regions. It is an occasional visitor to our shores, but is less often seen there than is the case with the Greenland species.

No. X.—THE SEAL TRIBE.

LIKE the whales, which formed the subject of the preceding paper of this series, the Seals are inhabitants of the water, although they are not entirely aquatic in their habits, as we shall presently see.

Supposing that we possessed no previous information upon the subject, a single glance at their outward structure would be sufficient to indicate to us the mode of life for which these animals were intended ; for the long, slender body, and the broad, webbed limbs, formed almost like the fins of a fish, would at once inform us that the water was their appointed habitation.

For an aquatic life the seals are singularly adapted.

The form of the body allows them to pass through the water with great facility, and the swimming powers are so highly developed that the animals are enabled to pursue and capture the fish in their own element.

If the motions of a captive seal are watched, as the animal disports itself in its tank, it will appear to pass through the water without active exertion, the lithe, supple body going through its manifold evolutions without any visible means of propulsion. By looking a little more closely into the matter, however, we may find out the secret.

The fore-limbs have but little to do with the matter,

being used as an aid to progression more upon dry land than in the water. Indeed, when the seal swims, it presses the fore-feet, or flippers, as they are called, against the body. The hinder feet, however, which are not rudimentary as in the whales, but are flat and broad, serve both to propel the body through the water, and also as a rudder to direct its course. And this is managed in a manner very different to what we might expect. The feet do not beat the water, like those of the duck or frog, and thus answer the purpose of oars, but are placed side by side, thus forming a single paddle, set edgewise, like the tail of a fish. These united feet being then swept from side to side, the animal is driven along by the action against the water, a slight alteration only of their position being required in order to direct the course.

By the aid of this simple means of progression, the seal is enabled to traverse the water with wonderful speed, and at the same time with an easy and undulating grace which is equalled by no other creature. This peculiar motion is owing in a great measure to the flexibility of the spine, which allows the body to be bent with facility in almost any direction.

Once upon dry land, however, the graceful movements degenerate into an ungainly shuffle, and the animal reminds the observer of some huge and overgrown caterpillar crawling awkwardly along. The fore-limbs now come into play, and the animal scuttles along, clumsily enough, it is true, but yet with considerable speed.

The nostrils and ears are provided with the means of preventing the ingress of water when the animal dives below the surface, the former closing by their own elasticity, while the latter are furnished with a structure analogous to that found in the whales. There is a curious point concerning the locality of the external ears. The orifice is not placed as we might think, directly over the organ itself, but below, and rather behind the eyes—a passage running beneath the skin to the ear itself.

The whiskers of the seal bear a very strong resemblance to those of the various animals of the cat tribe, their bases being connected with sensitive nerves which warn the animal of the slightest touch. It is thought that these may be of service in the pursuit of prey.

In order to protect the animal from the evil results of continued immersion, the body is clothed in a manner which effectually retains the vital heat. This is done in a threefold manner.

Beneath the skin is found a thick layer of fat, answering the same purpose as the blubber in the whales, and which of itself would, in water of ordinary temperature, be sufficient to retain the animal warmth. Inhabiting, however, as does the seal, the icy waters surrounding the polar regions, something more is necessary. This we find in a double coating of fur, the inner layer—the sealskin of commerce—being of a very fine and silky nature, and the outer of a more coarse and bristly character, serving as a kind of thatch to the whole. When the animal is in the water this fur is pressed closely against the skin, and, being constantly lubricated with an oily fluid secreted by the skin, is rendered perfectly water-tight, just as is the proverbial 'duck's back.' This coating of fat also serves another and very remarkable purpose.

During the breeding season the adult males come ashore, and each takes possession of a piece of ground, and occupies it together with a number of females. Each male endeavours to enlarge his territory so as to accommodate more wives, but those who occupy the adjoining spots will fight to the death rather than allow him to encroach on their property.

No male seal, therefore, can leave his ground for a moment, as, if he did so, the surrounding proprietors would at once take possession of it, and carry off his wives. Consequently, he cannot venture into the sea for food; and, if he had not some other mode of sustaining life, would soon die of hunger. But, just as the camel can sustain life by absorbing the fat contained in its hump, so the seal can feed—if we may use the term—on the thick coating of fat which envelops the body.

An adult male seal is really a formidable antagonist. When in best condition, it will weigh some twenty-five stone, and by the mere rush of its onset, and weight of its body, will overset any human foe who does not know how to oppose it.

Strong though the seal may be, it has one part of its body which is as vulnerable as the heel of Achilles. It has been already mentioned that a great number of nerves converge upon the upper lips and nostrils. A blow upon the nose will instantly stun, even if it should not kill, the most powerful seal that ever lived. But it must be well aimed, as the fatal area is very small; and if it be missed, the antagonist will find himself sprawling on the ground, and in very evil condition from the mere weight of the seal as it flounders over him.

My readers may remember the celebrated fight between Hector McIntyre and the ' Phoca,' in Sir W. Scott's ' Antiquary,' and the discomfiture of the too impetuous Highlander.

The food of the seal consists chiefly of fish, which its wonderful powers of swimming enable it to capture with little difficulty. Various molluscs and crustacea, however, also form part of its diet.

In order to assist the animal in seizing its prey, and also in retaining it when once secured, the teeth are formed in a very singular manner. The sharp canines are long and powerful, and the molars are covered with sharp projections of various sizes, so that even the most slippery fish, when once fairly seized, has very little chance of escape. The tongue, for some unexplained reason, is slightly cleft at the tip.

The young of the seal are very few in number, seldom being more than two, and generally only one at a birth. When newly born they are almost white, the colour gradually deepening as they advance in age.

For the first few weeks of their lives the young seals are brought up upon dry land, the mother carefully tending them until they are sufficiently strong to take to an aquatic life.

All the seals are inhabitants of the colder seas, and are especially numerous on the borders of the Polar Circle. The severity of the climate matters but little to them, for they are rendered secure from cold by their three-fold protection, and can always obtain prey beneath the ice, no matter what may be the season of the year.

In order to prevent the surface of the water from being completely frozen over, and separating them from the outside air, a number of seals are accustomed to congregate together into a single spot, thus keeping open a passage by the warmth of their bodies.

These passages are often utilized by the mother for the reception of her young, in the following manner :—

Ascending to the surface of the ice, the seal scrapes away the snow above the entrance to the passage, until she has excavated a small dome-like chamber, much wider than the passage itself. It will be seen that a ledge or shelf of ice must necessarily be left surrounding the aperture. Upon this shelf the baby seal is placed, and is there nurtured in safety by its parent until it is able to shift for itself.

Upon our northern shores, the Common Seal (*Phoca vitulina*) may mostly be seen, and on many parts of the Scottish coast is found in considerable abundance.

Seal-shooting is on these shores a sport which is quite as exciting as deer-stalking, and needs as much of the huntsman's craft. The seal is not only one of the most wary of animals, but it never shows more than its head above the surface of the water, and therefore affords a much smaller mark than a stag. Moreover, even when the sportsman has succeeded in hitting a seal, it often sinks as soon as hit, and he loses it.

Such a misfortune as a seal lost from sinking never occurs to the old hunter, and is a sure proof of inexperience. When a seal comes to the surface after a long dive it remains quiet for a time. A young hunter cannot resist so good a mark, sends a bullet through the animal's brain, and when he reaches the spot finds that it has sunk beyond his reach.

The reason is simple enough.

Like the whale tribes, the seals are enabled to remain for a considerable time under water in consequence of the power of aerating more blood than is required at the time.

Before diving the seal takes a number of long breaths, in order to aerate the blood fully, and just as it dives fills the lungs with a powerful inspiration. As it traverses its course below the water it allows the air to escape gradually, and its course can be easily traced by watching for the bubbles as they ascend to the surface.

Consequently, when the seal rises its lungs are empty, and if it be shot before it has had time to fill them it is sure to sink.

I

But if the hunter will have patience to wait until the animal has taken in its supply of air, and then shoot it, he will find that the air in the lungs will act like a float, and keep the body at the surface.

THERE are many different species of seals, several of which are named after animals which they are supposed to resemble. One of these is the Leopard Seal (*Leptonyx Weddellii*), or Sea Leopard, as it is indifferently termed, which derives its popular title from the whitish spots which are irregularly sprinkled over the body. This is not one of the larger seals, seldom exceeding ten feet in length, and not often attaining even to those dimensions. It is an inhabitant of various parts of the southern hemisphere.

A MORE interesting and well-known species is the Crested Seal (*Stemmatopus cristatus*), which is found upon the northern shores of America.

Upon looking at a specimen of this animal, the attention is at once struck by the remarkable crest, from which it takes its name. This strange structure springs from the muzzle, and rises to a height of several inches, supporting a kind of cowl which entirely covers the head. This curious organ, which is found in the adult male animal alone, is found to be a development of the 'septum,' or dividing gristle of the nose. As regards its object we are entirely at fault, the theory that it is intended to aid in the sense of smell at once falling to the ground when we consider that it is possessed by the adult males alone, being found only in a rudimentary form in the female and the young of both sexes.

Although the object of this strange development has not as yet been discovered, it is certainly of service to the animal as a means of protection when attacked by man. All the seals being particularly sensitive in the region of the nostrils, it is the usual practice of the hunters to stun them by a heavy blow upon the head, returning to complete the operation when the chase is over. The head of the crested seal is, however, guarded in a great measure by the curious helmet, and a blow sufficient to kill any ordinary seal serves only to stun the animal for a very short time.

When once roused, the crested seal is an active and formidable enemy, using both teeth and claws with considerable activity and address. Its strength, too, is very great, and as the animal averages some eleven or twelve feet in length, it will be seen that it is by no means a despicable foe.

The colour of the fur is a dark bluish black, paling almost to white beneath the body, and sprinkled with a number of greyish patches, each of which encloses a black spot. The head, tail, and feet are black. The fur is of considerable value in commerce, the skins being imported in great numbers.

Next comes the Harp Seal, or Atak (*Phoca Græn-landica*), a closely allied species.

This animal derives its somewhat peculiar name from the markings of the body, which are disposed in a very singular manner. The ground colour of the fur is a whitish grey, upon which are drawn two broad bars of a jetty black running from the shoulders, where they almost join one another, to the root of the tail. The form of these markings has been supposed to resemble an ancient harp; thence its popular title. The greater part of the head is also black.

This peculiar marking does not show itself until the fifth year of the animal's existence, the fur until then changing its colour and markings with every successive season.

The harp seal is an inhabitant of the coast of Greenland and Iceland, where it is found in great profusion, and is much sought after on account of the valuable oil which is obtained from the bodies. Some idea of the extraordinary abundance of these animals may be gleaned from the statement, made a short time since by one of the leading daily newspapers, that one vessel alone, in the course of a single voyage, had procured no less than fifty thousand carcases, valued at more than thirty thousand pounds. In the season of 1857, it is also said that half a million of seals were captured by the combined efforts of the vessels engaged in the trade. Success so great, however, is the exception, and not the rule, an ordinary season producing barely half the number of carcases.

Like the common seal, the harp seal is easily domesticated, being often tamed and taught to perform a variety of tricks. Both animals are remarkably docile and intelligent in disposition.

WE now come to the huge and ungainly monster known by the various titles of Walrus, Morse, and Sea Horse, and scientifically as *Trichecus Rosmarus*.

This is, perhaps, one of the most extraordinary of all the mammals inhabiting the water, its huge size, its bristle-fringed jaws, and the enormous projecting canine teeth, often nearly two feet in length, causing it to assume a strangely grotesque appearance.

In consequence of the size of these tusks, the jaw is much enlarged in front, the protuberant muzzle giving to the animal a very ferocious aspect. The nostrils, for the same reason, are placed very high in the head. The lower jaw narrows rapidly towards the centre, in order to pass between the two canine teeth.

The walrus is found in great quantities upon the borders of the Polar regions, both northern and southern, generally congregating in herds of some six or seven thousand in number. It is an animal of some value, even to civilised mankind, the oil and skin being in considerable request, while the tusks furnish valuable ivory, which has the advantage of retaining its colour for a very long period.

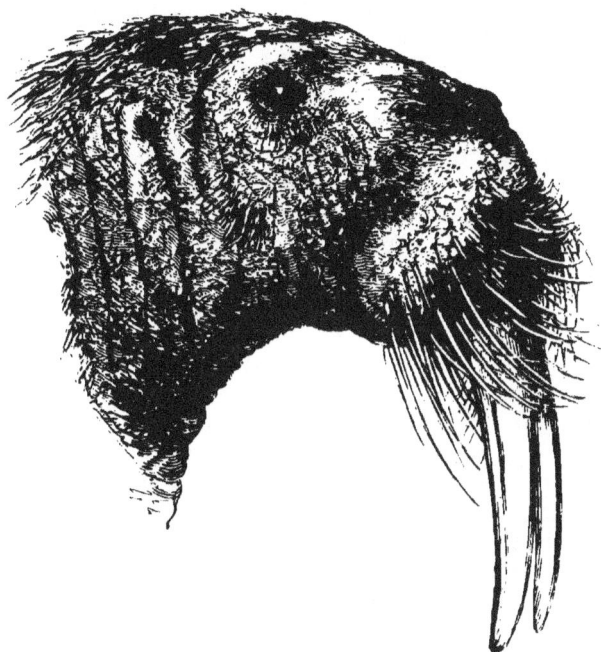

Head of Walrus.

To less civilised nations, too, the walrus is almost a necessity of life ; nearly all the soft parts of its body serve as food, while the intestines provide material for the construction of fishing nets, hooks and other articles being made from the tusks. The sinews serve as fishing lines, and the skin as a covering for the 'kajak,' or boat, so that from the animal are procured the whole of the paraphernalia for obtaining fish. The oil is burnt in the lamps, without which life would be an impossibility, and, in fact, the walrus is to these people what the camel is to the Arab, or the bison to the North-American Indian, a necessary adjunct to existence.

Only a short time before these lines were written, an entire tribe of Esquimaux perished because the walrus had deserted their coasts.

So needful do these tribes consider the walrus, that it is almost impossible to persuade them that life in any form is possible without the animals. They even carry the idea to a further pitch, and refuse to believe in the possibility of a future state of existence unless plenty of walrus are to be procurable. In vain the missionaries tried every means in their power to convince them into a belief in the Christian religion. ' No walrus, no heaven,' was the terse and conclusive answer, and there the matter had to rest.

Man is not the only being who recognises the merits of the walrus, for the Polar bear is fully aware of the advantages accruing to the fortunate slayer of the animal. He attacks it in a singular but very effectual manner.

Creeping closely up to the unconscious monster as it lies sleeping upon the shore, he leaps suddenly upon its back, clings to it with his hind feet and one of his fore feet, and delivers a series of tremendous blows upon its head with his unoccupied paw. Usually, this plan succeeds in a very short time, the mighty strokes first stunning the walrus, and at last smashing the skull.

Now and then, however, with an old and exceptionally thick-headed animal, the tables are reversed, the walrus plunging into the sea and carrying its opponent beneath the surface, where, in a few moments, he is obliged to relinquish his hold in order to obtain a supply of air, while his hoped-for prey makes good its escape.

The chase of the walrus, formerly a very easy matter before the animals had profited by bitter

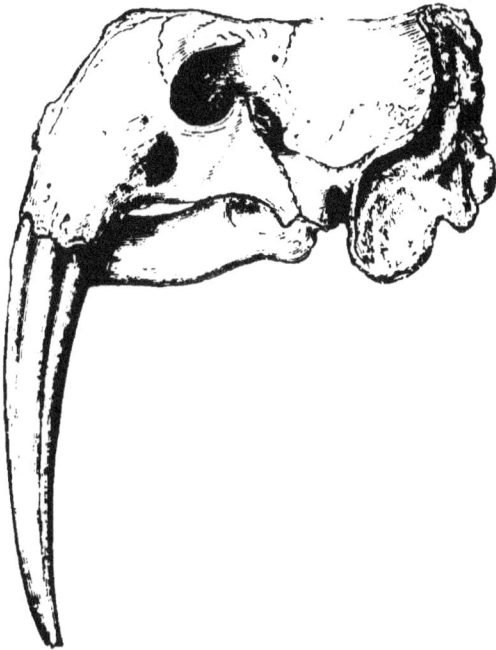

Skull of Walrus.

experience, is now a business requiring great caution and address.

The great difficulty is to prevent the animals from reaching the water, for which they at once make when alarmed, passing over the ground at a wonderfully rapid pace. As nothing can stand against the combined onset of the alarmed animals, dogs are employed in order to separate them as much as possible, and to distract their attention from their pursuers. When driven to bay the walrus becomes a very formidable enemy, the huge tusks being capable of inflicting most terrible wounds.

In spite of its enormous size, a full-grown walrus attaining to the length of fourteen or fifteen feet, and weighing as much as a large elephant, the animal has often been tamed, and has even been taught to procure fish for its master, and to bring them to land untouched.

ANOTHER strange and grotesque animal of this tribe is the Sea Elephant, or Elephant Seal (*Morunga proboscidea*), so called from its enormous bulk, and also from the elongation of the snout, which somewhat resembles the trunk of the elephant.

Huge as is the walrus, the elephant seal is far superior in size, having been known to reach the extraordinary dimensions of thirty feet in length, and eighteen feet in the circumference of the body. The strange nasal development, like the helmet of the crested seal, is found only in the adult males. Unless the animal is frightened or alarmed it is not very conspicuous; in such a case, however, the snout is thrown forward, and the animal blows through it with great violence, causing a strange harsh sound, which is audible at a considerable distance.

Notwithstanding their huge dimensions, the elephant seals are by no means so formidable as the walrus, apparently possessing no idea of revenging themselves upon their pursuers. When alarmed, they make for the water, the whole body quivering like so much jelly, on account of the fat with which the body is laden.

From this fat a valuable oil is procured in large

Sea Elephants.

quantities, a single animal yielding from seventy to eighty gallons; the skins, also, are of very fine quality.

The elephant seal is an inhabitant of the seas of the southern hemisphere, chiefly between 35 degs. and 55 degs. of south latitude. It is a migratory animal, travelling southwards as summer comes on, and northwards again at the approach of winter.

The last of the seal tribe which we can mention in this paper is the Sea Lion (*Otaria jubata*), an inhabitant of Kamtschatka and the Kurile Islands, and also of Northern America.

The dimensions of this animal are about equal to those of the walrus, fifteen feet being an average length. Upon the neck and shoulders is a thick mass of stiff bristles, somewhat resembling the mane of a lion; to this is owing the popular name of the animal.

The sea lion is remarkable for the hoarse roar which it continually utters when on land, the united clamour of a herd of these creatures being almost deafening to the human ear.

The disposition of the sea lion is very quiet and peaceable, the animal falling an easy prey to the hunters. Amongst the animals related to its own species, however, it is very fierce and determined, ruling supreme in its own dominions, and holding the smaller animals in complete thraldom.

It seems strange that so mild and apathetic an animal upon shore should be so tyrannous in the water. Yet such is the case, and the sea lion may fairly be considered, even the crested seal not excepted, as the most fierce and savage of all the seal tribe.

No. XI.—THE ELEPHANT.

THE Elephant belongs to the curious family of the Pachyderms, or 'thick-skinned' animals, which also includes such apparently dissimilar creatures as the little Hyrax, or 'Coney' of the Scriptures, and the various swine.

At first sight, and judging by their outward forms, these animals do not appear to have very much in common, and it may seem somewhat strange that they should have been included in the same family. An examination, however, of the fossil animals of the group supplies the requisite links, and affords conclusive proof that all these animals really belong to one and the same division.

As is well known, there are two distinct species of elephant, the one inhabiting Africa, while the other is a native of various parts of the continent of Asia. As far as their chief characteristics are concerned, the two animals are sufficiently alike to allow of a single description sufficing for both.

The most singular part of the structure of the elephant is, of course, the proboscis, or 'trunk' as it is popularly called, upon which the very life of its owner depends. This wonderful organ is, in reality, merely a development of the nose and the upper lip, the nostrils running through its entire length. The extremity is furnished with a curious finger-like appendage, which is of so delicate a nature that it can pluck a single blade of grass if required.

The proboscis is formed alike for strength and flexibi ity, and is provided with the enormous number of fifty thousand distinct muscles, some running longitudinally along the proboscis, while others radiate from its centre.

Upon the proboscis the entire nourishment of the animal depends, and, were it deprived of that organ, starvation would inevitably ensue. The short and thick neck would prevent it from grazing, while the long tusks would hinder it from devouring the herbage which grew on a level with its body. Water, also, could no longer be obtained, and thirst and hunger combined would shortly end the sufferings of the mutilated animal.

In somewhat the same manner as the camel, the elephant possesses the faculty of storing up water in the interior of the body, and is, moreover, able to withdraw it when required by means of the trunk, and sprinkle it over the body, in order to cool the heated surface.

The method of drinking is somewhat peculiar. Inserting the tip of the trunk into the pool, the animal sucks a quantity into its cavities ; the proboscis is then reversed, the end placed in the mouth, and the fluid discharged down the throat.

In providing food, too, the trunk answers much the same purpose, first plucking the leaves, etc., and then placing them in the mouth.

The head and skull of the elephant are formed in a very curious manner, affording a most wonderful example of strength combined with lightness. The former quality, of course, is indispensable, the enormous weight of the tusks and proboscis necessitating the provision of large and powerful muscles, while the advantage of the latter is self-evident

In order to satisfy these demands, the bony plates which form the skull are separated from one another, and form a series of cells, each of which contains a number of smaller chambers, bearing, in fact, a considerable resemblance to a honeycomb. These cells are filled with a thick oily fluid.

In the midst of these cells lies the brain, which is remarkably small in comparison to the size of the animal, and is thus protected from the effects of the various concussions it would inevitably receive during the headlong rushes of its possessor.

In like manner, too, this structure protects the animal in a great measure from the bullets of the hunter, for, unless they should happen to enter by the eye, the ear, or the nostrils, the leaden missiles stand but little chance of reaching the brain, burying themselves in the cellular mass, and doing comparatively little damage. An elephant has been known to receive between twenty and thirty large balls in the head alone before finally succumbing to its wounds.

It is, however, a somewhat curious fact that the skulls of the African and the Asiatic elephant are not formed exactly in the same manner, a'though the curious cellular structure is found in both, so that a bullet, which to the one would cause little injury, would to the other be certain death.

For example. If a hunter stands in front of an Indian elephant, and sends a bullet into the spot where the proboscis joins the head, the animal falls dead without a struggle, the bullet having penetrated to the brain. But if he were to attack an African elephant in the same manner, the bullet would pass above the brain, and waste itself in the cells of the skull, only irritating, and not injuring the elephant.

The limbs of the elephant, also, are admirably adapted to sustain the immense weight of the animal.

Stout, and comparatively short, they are set perfectly upright like so many pillars ; the hinder pair, also, are without the elongated cannon-bone, so that the so-called ' knee-joint' is absent, while the real knee is very conspicuous.

The elephant is a far more active animal than might be supposed, judging by outward appearance, his speed when excited being almost equal to that of a fleet horse. On rocky and mountainous ground, too, he is perfectly at home, notwithstanding his bulky proportions, and will ascend and descend acclivities where a horse is utterly unable to gain a foothold.

His method of descending precipitous ground is very singular. Kneeling down, with the fore legs stretched out in front and the hinder ones bent backwards, he gradually lowers himself towards the ground, making use of every little inequality in the surface, or scraping a foothold with his hoof, should a convenient one not be otherwise attainable. If the declivity be very steep, he pursues a winding course, just as does a horse in ascending a hill.

The foot of the elephant is wonderfully suited to the work it has to perform. The hoof which encloses it is formed of a vast number of horny springs, similar to those found in the hoof of the horse, which protect the foot from any concussion against the ground, and enable the animal to move with surprising ease and silence. Notwithstanding the huge bulk of the body, the tread of an elephant is perfectly noiseless, and as the animal possesses the faculty of forcing its way through the thickest jungle without snapping even a twig, a hunter may be in close proximity to a herd of moving elephants, and yet not be in the slightest degree aware of their presence.

The toes, five of which are found in each foot, are

almost entirely encased in the hoof, and are only partly visible to outward inspection.

The tusks of the elephant vary according to the sex, age, and species of the animal, being most highly developed in the male of the African variety. They do not, as might be imagined, spring from the jaw itself, but, like the teeth of the bottle-nosed whale, proceed from a vascular formation found upon the gums.

The molar teeth appear to be formed of a number of smaller teeth, which are closely fastened together, so as to form a single large mass. These are set obliquely in the jaws, and are gradually worn away by constant use, fresh teeth taking their places as often as required. In this manner, an elephant may have seven or eight sets of teeth, each set increasing in size in proportion to the growth of the animal. The incisors, which are found in the upper jaw only, are long and projecting; the canines are altogether wanting. The size of the elephant is generally much exaggerated, even a large animal seldom exceeding ten feet in height, while the average is a foot or so less.

The elephants of both continents alike are almost invariably found in herds, varying considerably in point of number, which are always under the guidance of some old and experienced leader. They dwell in the thickest forests, being seldom found at any great distance from water.

During the drought of summer, of course, the smaller streams often run dry. In such cases, the reasoning powers of the animals come to their assistance and the elephants, as Sir Samuel Baker tells us, in his well-known work ‘Eight Years in Ceylon,’ ‘make use of their wonderful instinct by digging holes in the dry sand of the river's bed; this they perform with the horny toes of their fore feet, and frequently work to a depth of three feet before they discover the liquid treasure beneath.'

WE will now devote our attention to a description of the habits, etc., of each of the two species, taking the Asiatic animal first in succession.

This elephant may be at once distinguished from its African relative by the size and form of the ears. those organs being in the Asiatic animal proportioned to the dimensions of other parts of the body, whilst in the African variety they are of very great comparative size, almost meeting at the back of the head, and hanging considerably below the neck. An African hunter has been known to shelter himself under an elephant's ear during a storm, and to emerge perfectly dry when the storm had passed over. The head of the former, also, is elongated, and the forehead concave, while in the latter the very reverse is the case.

Another point of distinction, too, is found in the molar teeth, the enamel upon the surface of those of the Asiatic elephant being moulded into a number of narrow bands, while in the African species it is formed in a series of diamond-shaped folds.

In the Asiatic species, too, the tusks are found only in the males, and are not possessed by every member of even that sex. When present, they are, generally speaking, very much inferior both in size and quality to those of the African animal.

The Asiatic elephant is chiefly remarkable for its services to man when trained to do his bidding. In all work which necessitates the employment of great strength united with intelligence, the value of the animal is inestimable. In such operations, for instance, as piling logs, laying dams, or even building walls, elephants are largely employed, their enormous strength and quick judgment rendering them most valuable auxiliaries.

These advantages are, however, not unmixed with drawbacks. The health of the animal, for example, requires constant care and attention ; the skin, thick as it is, is liable to abrasions, resulting in ulcerous sores ; and the eyes are constantly subject to inflammation.

Taking these disadvantages into consideration, many writers are of opinion that the value of the elephant as a beast of burden is greatly over-estimated, and that he is in reality of little more use than a powerful dray horse, which can work for longer hours, and is not so subject to sores and inflammations.

As far as some of the operations performed by elephants are concerned, this theory may be true enough, but when we consider the enormous weights which these animals are accustomed to carry, and the precision with which the largest and heaviest beams are placed by them, it seems hardly possible that their duties could be adequately performed by any other animal, no matter how powerful and intelligent it might be.

Another of the manifold purposes for which the Asiatic elephant is employed is that of an auxiliary in the chase of the tiger.

In order to serve in this capacity, the animals are captured when very young, and are carefully trained to perform their future duties. This is no easy task, for in the very nature of the elephant there appears to be an ingrained dread and abhorrence of the tiger, causing it to fly in terror from the mere sight of the skin of the fierce beast.

The education, therefore, of the hunting elephant is a matter of care, time, and patience, and is conducted as follows.

K

A tiger-skin is procured and stuffed, in order to resemble as far as possible the form of the living animal. This is continually presented to the elephant until he loses his natural fear of the striped skin.

The next step is to teach the animal to gore his foe with his tusks, and trample him under foot.

Next, a boy is placed inside the skin, in order to counterfeit the motions of the living tiger, and, finally, a dead animal is substituted for the stuffed skin.

Yet, with all the preliminary training, the elephant is seldom to be depended upon in the hour of actual danger, the rush of the furious tiger often causing the huge animal to turn tail, and fly before the onslaught of its foe.

During these expeditions, the animal is guided by a driver, or 'mahout,' who sits astride upon the neck, directing his charge by means of a spiked hook, or 'haunkus,' which is placed against the head of the elephant in such a manner as to convey the driver's instructions to the animal. The hunters ride in a 'howdah,' or car, which is fastened upon the elephant's back.

The elephants intended for domestication are cap tured in two ways. In the first of these, 'koomkies,' or trained female elephants are employed, which divert the attention of the intended captive from his approaching foes, who even creep beneath his body without alarming him, and place nooses of strong rope round his limbs. The ropes are then fastened to convenient trees, the koomkies called off, and the elephant finds himself a prisoner. For a time he struggles to release himself from his bonds, but finally yields to his captors, and is led off to a place of security.

The second method of capturing the elephants is of a far more comprehensive nature, all the members of one or more herds being included in the attack.

For this purpose a large enclosure, or 'keddah,' is formed of stout posts, which are driven into the ground at such a distance from one another as to allow a man to pass freely between them. A head of elephants is then surrounded by hunters, and gradually driven towards the keddah, the door of which is left open.

By slow degrees, the operation sometimes extending over several weeks, the animals are forced into the enclosure, the entrance to which is immediately closed.

Should the animals attempt to burst from their place of confinement, they are immediately driven back by torch-bearers, who thrust their flaming brands into the faces of the excited captives, and deter them from breaking through the walls. After a time, the imprisoned animals relax their efforts to escape, when the hunters cautiously enter, and bind each of them securely to a tree, or other immovable object.

The nature of the Asiatic elephant is, as a rule, very quiet and peaceable, forming a great contrast to the fierce and savage character of its African relative. Even when hunted, if it should be successful enough to strike down its foe, the animal seems to have little idea of revenge, and usually contents itself with kicking its prostrate adversary from foot to foot without causing any great injury.

It may seem remarkable that a domesticated animal should be desirous of reducing its fellows to a state of servitude. Yet the elephant does so, the females using every means in their power to capture the males.

One case is known where a female escaped from her owners, carrying with her a chain. In a few days she returned, and by signs and sounds told her keepers that she wished them to accompany her into the

forest. This was done, and she led them to a spot
where a fine male elephant was found chained to
a tree.   In fact, she had acted the part of Delilah
towards Samson.

THE African elephant (*Elephas Africanus*) is spread
over a large tract of country, extending from Abys-
sinia to the borders of Cape Colony.   Like the Asiatic
species, it is an inhabitant of the thick forests, seldom
venturing into the open country.

This elephant is also much sought after, although
from very different motives to those which influence
the hunter of the Asiatic animal.

The natives of Africa are either not aware of the
services rendered by the elephant when captured and
carefully trained, or mingled apathy and fear prevent
them from availing themselves of their opportunities.
Just the same is the case with the Chetah, or Hunting
Cat, which in Asia is carefully trained for purposes of
the chase, while in Africa it is allowed to remain in
freedom.   Formerly, however, the African elephant
was trained for purposes both of war and peace, just as
is now the case in India.

The ivory of the tusks forms the principal incentive
to the efforts of the hunters engaged in the chase of
this animal, being of very fine quality and considerable
value.   An ordinary pair of tusks, weighing, perhaps,
rather over a hundredweight, will fetch thirty-five or
forty pounds, although the price varies slightly according to the condition of the market.

The flesh, too, is by no means an unimportant
article of diet, especially among the natives, to whom
the slaughter of an elephant is an occasion of great
rejoicing.   Some parts, such as the foot, are justly
considered as especial dainties, but the greater portion
of the flesh is stated by many travellers to be little

superior in toughness and flavour to ordinary shoe-leather.

The foot is baked in a somewhat curious fashion. A fire is lighted upon the ground, and allowed to burn itself out. A hole is then dug beneath the spot, and the foot is inserted, being then covered up with the warm earth. A second fire is now lighted, which is also suffered to burn itself out, and when the earth is thoroughly cool, the process is complete, and the dainty in perfect order for the table.

Until the advent of firearms, the slaughter of an elephant was only a very occasional event with the natives, who were either obliged to follow it for days, attacking it with their spears at every opportunity, until the animal fell from sheer exhaustion and loss of blood, or to trap it by means of pitfalls.

These latter are still employed, being dug in the paths of the animals, and covered over with boughs and earth to imitate the surrounding surface. With the old and experienced leaders, however, these precautions are of little avail, for the crafty animals test every inch of ground with their trunks before trusting their weight upon it.

Should one of the animals, however, be unfortunate enough to fall into the snare, it has no chance whatever of escape, a sharp upright stake being fixed in the centre of the pit, upon which the luckless creature is impaled by its own weight.

In their wild and free state, it is probable that elephants live to a very great age, and even when domesticated, appear to be long-lived animals ; there have been several apparently well-authenticated instances of these animals attaining the age of two hundred years.

No. XII.—THE RHINOCEROS, HYRAX, AND HIPPOPOTAMUS.

THE RHINOCEROS is connected with the elephant by a number of links, such as the Tapir, in which a small and imperfect proboscis is present, and the various swine, in which the proboscis is modified into a very mobile, blunt-tipped snout, with the nostrils at the extremity. Geology, too, has revealed traces of many animals which are now extinct upon the earth, and which render the transition between these animals very much less abrupt, conclusively proving their approximate position in the scale of creation.

The rhinoceros, of which several species are known, is found in various parts of the African and Asiatic continents, preferring those neighbourhoods in which water is easily to be obtained. Although the various species differ in several minor characteristics, they are sufficiently alike in their chief peculiarities of structure to allow of a single description sufficing for the whole.

The so-called 'horn' is naturally the first point to attract our attention; and a very curious and wonderful object it is.

Notwithstanding the powerful shocks it is called upon to bear, and its uses as a weapon of offence, it is not in any way connected with the skull, as is almost universally imagined to be the case. It is, in fact, merely a growth from the skin, from which it may be removed by a few cuts around the base from a keen-bladed knife. An ordinary penknife is quite sufficient for this purpose. This horn must be ranked in the same category with hair, spines, and quills, the structure being precisely similar in all. This may be at once proved by an examination of the horn, which, although smooth and polished at the tip, is separated at the base into a number of filaments, the hair-like formation of which may be easily seen.

In order to avoid the effect upon the brain of the violent concussion caused by the headlong charges of the animal, the bones of the face are modified in a very remarkable manner, forming a kind of broad and strong arch, one end of which is left free and unsupported. Above this end the horn is situated, the elasticity of the bony arch effectually breaking the force of the shocks. The horn does not attain its full dimensions for several years after the birth of the animal.

In olden times, the horn of the rhinoceros was held in great estimation on account of its supposed poison-detecting powers, and bore a fancy value in consequence. Eastern monarchs, for example, were accustomed to have their drinking-cups formed from the horn, the superstition being that any poison introduced into the vessel would cause the contents to bubble violently, and so bring about a discovery of the attempted assassination.

At the present time the horn is still of considerable value, being largely employed in the manufacture of umbrella handles and various other articles.

The skin of the rhinoceros is of great thickness, and of so tough a nature that it will resist any but a specially hardened bullet. The balls used in the chase of the animal are therefore alloyed with solder or tin or mercury, in order to supply the requisite hardness.

By the natives of both Africa and Asia the skin of the rhinoceros is greatly prized, being largely utilised in the manufacture of shields, which form a most effectual protection from spears, no matter how keen their points or how great the force with which they are hurled. Even a rifle bullet, indeed, unless fired at close quarters, would probably be checked or turned aside in its flight.

Yet, stout as is the skin in most parts of the body, there are places where its character seems to be altogether changed. In the Asiatic species of rhinoceros, for instance, the skin falls in heavy folds upon the neck, shoulders, and flanks, forming flaps which may be lifted up with the hand. Beneath these folds the skin is of a much softer and more delicate nature, and may be pierced without any very great difficulty. This fact is taken advantage of by the various parasites which infest the tropical forests, and which insinuate themselves beneath these folds, directing their attacks upon the thinner skin lying beneath them, and driving the animals almost mad by their incessant persecutions. Upon the under side of the body, also, the skin is comparatively soft.

Here we find a reason for the fondness of the rhinoceros and its allies for wallowing in the mud, the thick tenacious substance rapidly hardening beneath the rays of the sun, and affording an impenetrable barrier to the tiny assailants.

The eyes of the rhinoceros are by no means large, and are placed rather deeply in the head, the sight consequently being of a rather imperfect nature; in fact, the animal is unable to see any object directly in its front. The senses of scent and hearing are, however, developed to a considerable extent, and fully compensate the animal for its partial lack of visual power.

The INDIAN RHINOCEROS (*Rhinoceros unicornis*) is chiefly remarkable for the comparatively small size of the horn, the height of which sometimes little exceeds the diameter. It nevertheless forms a most effectual weapon, a well-known traveller stating that this animal is able to hold its own against an adult male elephant.

Another of the Asiatic species is the Sumatran Rhinoceros, which is provided with two horns upon the head instead of one. It does not appear, however, to make use of its formidable weapons, for its disposition is very quiet and timid, the animal flying from the presence of danger, and seldom facing even a single dog.

FOUR distinct species of rhinoceros are at present known to inhabit Africa, and it is yet uncertain whether still others do not exist.

The best known of these is the Rhinaster, Borele, or Little Black Rhinoceros of Southern Africa (*Rhinoceros bicornis*), which may be known by the shape of the horns and the upper lip.

The anterior horn is long, pointed, and curved backwards towards the head, while the posterior one is small and conical, closely resembling the weapon of the Indian rhinoceros. The upper lip, which is sharply pointed, overlaps the lower to a considerable extent.

The Borele is usually considered to be by far the most savage of all the species of rhinoceros, and the natives are said to fear the animal far more than they do the most infuriated lion. When wounded it is a truly dangerous opponent, and will attack its foe with a ferocity and determination which render escape a matter of considerable difficulty.

During the day-time the animal is seldom to be seen, selecting some secluded retreat in the thickest part of the forest, and there passing the hours of daylight. When night sets in, however, he awakes, and at once sets out for the nearest pool, in order to slake his thirst before prosecuting his search for food.

This he generally finds in various roots, which he ploughs out of the ground by means of the powerful horns, and also in the young shoots of the ' wait-a-bit ' thorn. Clumsy as it is in appearance, this rhinoceros is yet active and agile to a wonderful degree, possessing considerable speed, and severely trying the powers even of a good horse when attempting to escape from its furious onslaught.

Another well-known African species is the Keitloa or Sloan's Rhinoceros (*Rhinoceros keitloa*), which may be readily distinguished from the borele by its horns, which are of considerable and almost equal length. It is altogether a larger animal than the preceding, and is, if anything, even more to be dreaded as a foe, owing to its superior strength and length of horn.

Both the borele and the keitloa are black in colour ; there are, however, two African species of rhinoceros in which the colour of the skin is a greyish white.

The first and more abundant of these is the common White Rhinoceros, or Muchuco, as it is termed by the natives (*Rhinoceros simus*), which differs considerably in appearance from the two above-described species. Setting upon one side the colour of the skin, the chief differences may be summed up as follows. The muzzle is square instead of pointed, the head is elongated, and the anterior horn attains to considerable dimensions, three feet being by no means an uncommon length. The second horn, however, is of far lesser size, and closely resembles that of the borele.

In disposition, also, the animal is very different, being as mild and peaceable as the borele and the keitloa are fierce and savage. Even when attacked it seldom assumes the offensive, but generally seeks safety in flight instead of endeavouring to revenge itself upon its pursuer. Should its young be assailed, however, it will fight with great fury, and is then to the full as dangerous an opponent as either of its relatives.

The second of the white species, viz., the Kobaoba, or Long-horned White Rhinocer s (*Rhinoceros Oswellii*), is a very much rarer animal, and is far less generally distributed

The anterior horn of the kobaoba is of considerable size, sometimes exceeding four feet in length. Owing to the manner in which the head is carried, this horn, which is almost straight, and is directed forward instead of backward, is generally found to be more or less worn away by the friction with the ground. In consequence of its length and straightness the horn is of great value in the market.

A walking-stick made of a single piece of this horn will fetch almost any price in London or Paris. In the old days of muzzle-loading rifles, a ramrod made of rhinoceros horn was invaluable, as it was almost unbreakable, and yet was tolerably light to carry. A large knob was left at one end, and so it became not only a loading rod, but a formidable weapon. Short clubs of similar form are much used by the Kaffir tribes in hunting, and are called knob-kerries.

All the African species of rhinoceros are occasion-ally to be seen in small herds of eight or ten specimens, but can yet be scarcely described as gregarious, each animal in time of danger separating from his companions and selecting his own path. They are not prolific animals, a single young one only being produced at a birth.

THE HYRAX.

We are told in Ps. civ. v. 18, that 'the rocks are a refuge for the conies,' and in Prov. xxx. v. 26, reference is made to the same animal. 'The conies are but a feeble folk, yet make they their houses in the rocks.' The 'coney' is also named in the book of Leviticus as one of the animals which might not be eaten by the Jews.

Now, the coney which is here mentioned is not the rabbit, as most readers of the Scriptures suppose.

Rabbits are not frequenters of rocks. They live in holes which they excavate with their fore-feet, so that they need a tolerably loose soil, and would be entirely at a loss among hard rocks. The animal in question is a little creature, so like a rabbit in general appearance that it might well be mistaken for that animal.

Its teeth are apparently those of a rodent, and its feet look very much like those of the rabbit. It is clothed with brown fur, very much like that of the wild rabbit, and is wonderfully active, darting about with such rapidity that the eye can scarcely follow its movements.

But, when examined by the eye of the zoologist, the rodent-like teeth are evidently those of a miniature hippopotamus, and the paw-like feet are seen to be composed of hoof-clad toes like those of the rhinoceros. The rhinoceros has three toes on each foot, while the hippopotamus has four. The Hyrax has four toes on the fore-feet, and three on the hind-feet. In fact, this little creature, so apparently dissimilar to the hippopotamus and rhinoceros, forms a connecting link between them.

The species of Hyrax which inhabits Africa is popularly called the Rock-rabbit by the English colonists, and Klip-das by the Dutch. Its scientific name is *Hyrax Capensis*, that of the animal mentioned in Scripture being *Hyrax Syriacus*. It is called 'Ashkoko' by the natives.

Both creatures have similar habits. They are exceedingly wary, darting into the recesses of the rock at the slightest alarm, or, if they fear that their movements may betray them, crouching motionless against the rock and resembling it so closely that the keenest eye can hardly detect them.

THE HIPPOPOTAMUS.

Having now traced the connection between the elephant and the swine through the tapir, and that between the rhinoceros and Hippopotamus through the hyrax, we come to the Hippopotamus itself.

This animal (*Hippopotamus amphibius*), often known as the River-Horse, or Sea-Cow, is a native of various parts of Africa, being never found very far from the neighbourhood of water. Huge as is the animal, its size lies chiefly in the bulk of the body, the legs being very short, and the actual height seldom exceeding five feet.

The teeth of the hippopotamus are of wonderful size and curious shape, the canines being strongly curved, while the incisors lie almost horizontally. These latter are chiefly used in tearing up the various aquatic plants upon which the animal feeds. These teeth are of very fine quality and close consistency, and, the ivory obtained from them retaining its colour for a great length of time, are of considerable value, averaging in price from £1 to £1 5s. per lb. A single tooth is usually from five to eight pounds in weight.

Formidable as these teeth appear, they are employed solely for the purpose of feeding, unless the animal is wounded, or otherwise irritated. For their legitimate purpose they are most suitable implements, capable of severing a stem of considerable size, or of cropping the herbage as closely as if a scythe had been employed.

In consequence of its huge appetite and destructive habits, the hippopotamus is an object of great detestation in the neighbourhood which it frequents, as its constant visits to the plantations in the v'cinity result in an almost total destruction of the crops. And the havoc it causes is all the greater on account of the position of its legs, which, being very short, and set widely apart from one another, oblige the animal to make two distinct tracks, thus exactly doubling the damage caused by its passage through the crops.

In order to check its ravages as much as possible, various means are employed, chief among them being the pit-fall and the ' down-fall.'

The first of these needs no explanation, and the latter may be described in a few words.

A log of wood is heavily weighted at one end, and furnished with a spear-head dipped in poison. This is suspended to a branch over the path of the animal. To it is fastened a cord, which is carried across the track in such a manner that the pressure caused by the advancing hippopotamus causes the log to fall, and the poisoned spear to sink deeply into its body. The doom of the animal is then effectually sealed, the venom performing its destructive work in the course of a very short time.

The hippopotamus is also slain by means of a specially constructed harpoon, consisting of a stout shaft, some ten or twelve feet in length, and a barbed point fitting loosely into a socket at the end of the shaft, to which it is fastened by means of a rope composed of a number of separate strands ; this is in order to prevent the animal from biting it asunder. To the handle is fastened a stout line, to the other end of which a float is attached.

When an attack upon one of these creatures is contemplated, the hunters proceed, by means of a raft, into the midst of a herd, and plunge the harpoon into the body of the nearest animal. The wounded hippopotamus immediately dives, but is unable to shake off the harpoon, owing to the barbed point. As often as he rises he is attacked with spears, etc., which speedily complete the work of destruction.

Sometimes the wounded animal turns savagely upon its pursuers, and succeeds in tearing the raft or boat to fragments, occasionally killing one or more of the crew before they are able to reach the land.

The hippopotamus is generally found in tolerably large herds, each consisting of from twenty to thirty or more animals. Chiefly aquatic in their habits, they are generally to be found in the larger rivers and lakes. They are not averse to salt water, being often noticed floating or swimming in the sea itself.

The hide is enormously thick and strong, being fully two inches in thickness along the back. The well-known 'sjambok' whips are formed from this skin, which is prepared for use in a somewhat curious manner. Strips of suitable length are cut, and are then beaten with a hammer in order to consolidate the substance of the hide. This done, they are thoroughly dried, and are finally rounded off and polished by means of a sharp knife and sandpaper.

A well-made sjambok is a terrible weapon, capable of cutting a deep groove in a deal board with a single well-directed stroke.

There are two kinds of sjambok. One is used for driving oxen, and is attached to a long bamboo handle in such a manner that the complete instrument looks like a gigantic fishing-rod. In the hands of an experienced driver it becomes a terrible weapon, to which even the tough hide and obstinately sluggish nature of the African draught ox are forced to suc-cumb.

The driver can direct his blow with unerring cer-tainty, and a single stroke will cut completely into the skins of two oxen at once, raising a cloud of hair, and often causing blood to spirt from the wound. The crack of this sjambok is as loud as a pistol shot, and at the very sound of it the oxen fling them-selves against the yokes so as to avert the dreaded blow.

The short sjambok is popularly known as the ' cow-hide,' because it is made of the hide of the ' sea-cow.'

In order to retain the vital heat below the surface of the water, a similar provision is found in the hippo-potamus to that with which the whale is furnished. Beneath the skin is a thick layer of fat, a wonderful non-conductor of heat, while a system of glands keep the outer skin constantly lubricated with oil, and thus prevent it from coming into actual contact with the watery element.

This sub-cutaneous fat-layer is considered a great delicacy, and is known by the Dutch colonists as ' Zee-koe speck,' or sea-cow bacon.

The animal possesses the remarkable power of sink-ing the whole of its body beneath the surface at will, and remaining under water for a considerable period of time. This feat is accomplished by the contrac-tion of the body, so that, while the weight remains the same, the bulk is considerably decreased, and a lesser quantity of water displaced.

Like the rhinoceros, the hippopotamus, whether upon land or in the water, is usually a very sluggish animal, and seems greatly averse to active exertion. When attacked, however, or otherwise alarmed, it exhibits the most wonderful activity, dashing through the water in a series of tremendous leaps, or rushing through the forest with the most unexpected rapidity. Few obstacles can stand against the onset of an infuriated hippopotamus, the great bulk of the animal bearing down almost any barrier that can be opposed to it.

The colour of the hippopotamus is a dark brown, marked with a number of irregular lines resembling the cracks on the surfaces of old oil paintings. The skin is marked with a number of sooty black blotches, which are only visible, however, upon a close examination. When the animal is living in its native land, the ears, nostrils, and the ridge over the protruding eyes are of a bright red colour, and form capital marks for the rifle of the hunter.

The slaughter of the hippopotamus by means of the rifle is not at all an easy process, as the animal, when alarmed, sinks at once to the bottom, and only occasionally rises to the surface for air. Even on these few occasions the nostrils alone are exposed above the surface, so that only a very well-directed bullet can do any harm.

The hunter always endeavours to lodge his first ball in the nostrils, as, if they be wounded, the animal is unable to remain submerged. A second bullet in the eye, or behind the shoulder, will then mostly complete the business.

When in her native rivers, the female hippopotamus is a model parent to her offspring, carrying her cub about on her back, and tending it with the most affectionate care and solicitude. In confinement, however, she generally behaves differently, and has on more than one occasion been known to kill her infant in a fit of passion.

In former days the hippopotamus was a native of Europe, and the fossil remains are found even in our own country, the London clay being especially prolific in these relics of a bygone time.

L.

No. XIII.—ELEPHANTIANA.

"THE fear of you and the dread of you shall be upon every beast of the earth, and upon every fowl of the air, upon all that moveth upon the earth, and upon all the fishes of the sea ; into your hands they are delivered " (Gen. ix. 2).

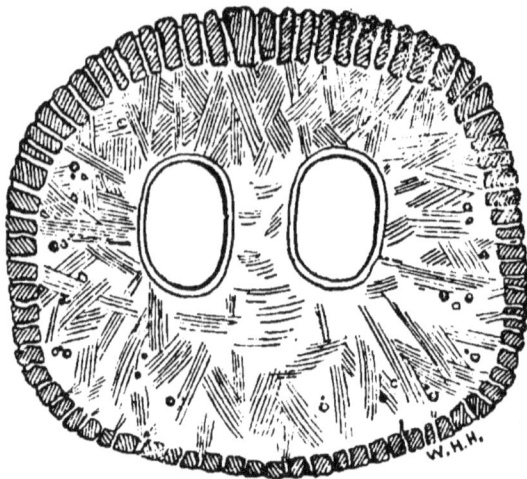

Section of Trunk of Elephant.

These bold and uncompromising words, written at least four thousand years ago, are absolutely true now, as they were then.

Wild animals, no matter what they may be, instinctively flee from man. The domesticated horse, which has never seen a beast of prey, trembles with terror at the smell of a distant menagerie; but the lion which inspired that terror is, in its wild state, quite as much afraid of the odour of man.

Let a lion but detect the dreaded emanation of man, and he slinks off as quickly as he can.

For, though we are happily unconscious of it, a very powerful odour emanates from all human beings, and strikes terror into wild animals. Deer-stalkers know well that they must approach a stag against the wind, for that even at the distance of a mile the stag can detect the presence of man, should the wind blow from him and not to him.

Similarly, the practical rat-catchers will never touch a trap with bare hands. They wear gloves rubbed with aniseed, and imbue the soles of their boots with the same perfume, before they can venture to handle a trap or to walk near the spot where the trap is set. Inexperienced persons neglect these precautions, and in consequence, the rat detects the dreaded odour of man, and keeps aloof from it.

Mole-catchers, again, always keep the skin of a dead mole by them, and rub it between their hands, before they set their trap, so as to overpower the natural odour of the hand.

Of course, there are some animals, such as lions, tigers, and the like, which will attack and devour human beings. But these are exceptional individuals, being almost invariably the aged animals, which have become too infirm to catch prey in the ordinary fashion, and are reduced to lurking about villages and pouncing upon any unwary straggler.

It is well known that the skin of a "man-eater," whether lion or tiger, is never worth preservation, being mangy, bald in patches, and altogether unsightly. Its skull is equally useless as a specimen, the teeth being blunt, worn down and decayed.

There is no animal, however gigantic, however fierce, however powerful, of which man is not the master. In proportion to his bulk, man is perhaps the weakest of living beings, and yet he is master of the strongest.

Not only can he destroy them—that is comparatively a simple task—but he can take them from their own savage life, and force them to become his servants.

So he has taken possession of the horse, the camel, and the ox, and made them bend their backs to the burden and submit their necks to the yoke.

He has reclaimed the dog from a predaceous life, and taught some of them to guard the flocks which in the wild state they would have devoured, and to be the friends and companions of their masters. Others he has taught to chase prey, not for themselves but for him.

He has taught the falcon to chase birds for him in the air, and the otter and cormorant to catch fish for him in the water. They not only do his work, but are proud of doing it, and contemptuously reject the society of their relatives who live only for themselves.

No better example of the universal mastery of man can be found than in the tame elephant. What is a man, that he should make the mighty elephant obey his orders? The creature could crush him in a moment, and in a fit of blind fury will do so. But when it is in its senses, the elephant acknowledges man as its master, and becomes his obedient servant.

Man rules by two means, Fear and Love. There are some beings which, from no fault of their own,

are so constituted that they must be made subject
to fear before they can learn to acknowledge Love ;
and this is the case, not only with different animals,
but with different individuals belonging to the same
species.

Take, for example, the dog. There are some dogs,
just as there are some men, which are constitutionally
ill-tempered, violent, ungrateful for kindness, mistaking forbearance for weakness, and ready to bite
the hand that feeds them. It is impossible to rule
them by love, until they have learned to fear, and can
understand that the hand which gives food can with-
hold it at will, can render them powerless at will, and
can at will inflict pain without the possibility of their
evading or avenging it.

Having, then, been taught by fear to acknowledge
that man is their master, they can begin to learn to
be grateful for their food, and to lick the hand which
gives, in lieu of biting it. For such dogs a severe
chastisement is really the kindest of lessons, and
although "force is no remedy," it is often a needful
preliminary before applying the remedy.

But there are dogs, as there are men, of a higher
order, which are absolutely amenable to Love, but
would be only made obstinate and resentful by
fear. Such an one was my bull-dog "Apollo."
Possessing all the concentrated strength and courage,
added to the instinctive combativeness of his race,
which make the thoroughbred bull-dog one of the
most wonderful animals in the world, he could be
compared to nothing but the Faure "accumulator,"
wherein a million foot pounds of force can be carried
in one hand. Despite his powers, which none knew
better than himself, he was one of the gentlest and
most obedient dogs that I have ever dealt with. I had
him when he was but a puppy, and never once beat

or scolded him.   Yet, he would obey the lifting of
my finger, or the glance of my eye, and the very idea
of incurring my displeasure was unendurable torment
to him.

All servants of the pen must of necessity be so
absorbed in the evolution of ideas and the balance of
sentences, that they are unconscious of time, space,
hunger, thirst, cold, or other material conditions.

It has happened that while I have been thus
absorbed, Apollo has tried to attract my attention,
and in failing, has taken it into his loving brain that
he must have offended me in some way.   On such
occasions he grovelled on the floor, he whined, he
licked my hand, and lay in abject despair until again
noticed.

So it is with elephants.   There are not two elephants with precisely the same disposition, and the
best keepers are those who try to find out the
peculiar disposition of the creatures under their care,
and to treat them in accordance with that disposition.

As elephants, like falcons, are seldom bred in
captivity, but are captured when wild, it necessarily
follows that the first lesson they must learn is to fear
man, and to realize the strange fact that he is their
master.

It is a remarkable fact that there is no task which
tame elephants undertake so willingly as that of
capturing their wild relatives.   They seem to enjoy it
with all their hearts; and both sexes are equally keen
at the sport, the females acting as decoys, and the
males as the representatives of force.

Supposing that a male has to be captured, two
female "koomkies," as these decoys are called,
saunter leisurely along, and soon make the acquaint-
ance of the victim. Each has her keeper, or "mahout,"
on her back, and it is a curious fact that an elephant

never seems to notice a man as long as he is on
another elephant's back.

The koomkies manage to place themselves on
either side of the male, and by degrees sidle him
close to a tree. One of the mahouts then slips
quietly to the ground, and while his koomkie and
her companion are distracting their victim's attention,
he passes strong cords round the animal's ankles, and
then makes them fast to a tree. The koomkies will
often aid him in this part of the work by taking the
ropes in their trunks, and passing them to their
master's hand as he wants them.

A similar process is then pursued with the forelegs,
and then the treacherous koomkies suddenly slip off,
leaving their dupe fast bound to the tree.

Sometimes, before the ropes are firmly tied, the
elephant becomes suspicious and tries to escape.
The koomkies employ all their blandishments to lull
his suspicions; but if he should still resist, and be
too strong for them, a powerful male is summoned
to their help, and all three beat him and hustle him
about until he is quite bewildered, and at last is held
firmly against the tree where the mahout is ready with
his ropes.

In either case, the duped elephant is left alone, his
vast strength paralyzed in some mysterious manner,
and his struggles for freedom only resulting in pain.
Those who have witnessed these struggles say that
the contortions into which a bound elephant will
fling its body are almost incredible. It rolls over and
over, it rises itself on its hind feet, butts at the tree
and tries to knock it down, uttering all the time its
screams of mingled terror and anger.

After a while comes another feeling. It can neither
eat nor drink, and the pangs of hunger and thirst are
felt. At first, it seems to resent the unwonted feeling,

but by degrees is so exhausted that it lies motionless on its side.

Now comes its future mahout. Bringing green food in his hand, he cautiously approaches the prostrate elephant. Mostly, rage will for the moment overcome hunger and weakness, and the animal will try to attack the man. In that case, the mahout quietly retires, and leaves the elephant for a few more hours.

This process is repeated until the elephant no longer tries to resist. He has learned his first lesson, that man, small as he may be, is, in some inexplicable manner, stronger than himself, and that resistance is useless.

Now he will take the grass from the mahout, and before long he welcomes the man's presence as the only mode by which he can obtain food. Sooner or later the lesson is learnt, and the captive acknowledges himself in subjection. The koomkies are again summoned, his hind feet are freed from the ropes, though the fore feet are kept hobbled, and, guided by the koomkies, he is taken to his future home, his new master seated on his neck.

As the late Mr. Rarey found to be the case with horses, the subdued elephant entirely trusts in the man who has conquered him, and even conceives a strong affection for his captor.

Sometimes the elephants, instead of being taken singly, are partly enticed and partly driven into a large and very strong enclosure, called a "keddah." This is made of massive posts planted deeply in the ground, and set far enough apart to allow a man to pass easily between them. These are supported on the outside by stout buttresses, so as to withstand the charge of the trapped elephants.

Fortunately, a herd of elephants never unites in a

charge on a given spot. If they were to do so it would be scarcely possible to build a keddah which could withstand them. As it is, the posts and buttresses need only be strong enough to resist charges of single elephants.

Once inside the keddah, the elephants are never allowed to rest. By means of fireworks, guns, torches, and shoutings, the elephants are driven backwards and forwards until they are fairly wearied out, and huddle together without even thinking of escape. Then come the hunters, with their koomkies and ropes, and bind the limbs of the wearied animals before they can understand what is happening to them.

Whether the elephant be taken singly or in numbers, the first lesson which it must learn is that it fears man as being stronger than itself, and that therefore it must obey him. Next, it learns to trust to man for food, and is not long before it learns to love him.

But, when, as was the case with the grand African elephant "Jumbo," the creature has lived with man from its infancy, the preliminary lessons are not needed, and man can rule the animal by love without any mixture of fear. On more than one occasion, when Jumbo was disposed to be rather wilful, his keeper, Scott, was urged to use his whip. This he invariably refused to do, saying, that if he were once to do so, his influence over the animal would be gone.

I fully believe that if I had even once used the whip to Apollo, his absolute belief in me as a being whose displeasure was infinitely worse torture than bodily pain, would have been lost. No creature can defy the extreme of bodily pain more heroically than a thoroughbred bull-dog. Diabolically cruel

experiments have been tried on the animal, and a
bull-dog has endured the severest tortures without
flinching. Pain he would not have feared, but he
did fear the loss of my love for him.

Not but that force may not be sometimes
necessary with any elephant. However gentle an
elephant may be, it is liable to occasional aberrations
of temper, which affect it much as a half-grown cat
is often affected with fits. The animal loses all control
over itself, and for a time is subject to raging madness.

Now, even a cat can do much harm during a fit,
and what a terrible creature a mad elephant must
be can be well imagined.

The elephant keepers of India, when they perceive symptoms of coming madness, fasten the
animal to a tree just as if it had been newly taken.
They put it on very low diet, and if it should be very
outrageous, they employ their largest and strongest
male elephants to assist in coercing it. These
animals understand the necessity of restraining their
companion, and if other means fail, will beat him
with their trunks when he tries to break his bonds.

Only a short time before these lines were written,
a remarkable instance of madness in an elephant
occurred in Siam.

In that country an albino, or, as it is generally
called, a White Elephant, is held to be, not a mere
animal, but a material habitation of Divinity, and is
honoured accordingly, even the king paying homage
to it. The White Elephant is addressed as "Sublime
Grandeur." He has his court and household officers
like the king. He is lodged in a palace, and is
decorated with jewels of priceless value. The
"Order of the White Elephant" is in Siam what the
Garter is in England, or the Golden Fleece in Spain.

A short time ago, one of these elephants was un-expectedly seized with madness.

He began by trampling to death five of his atten dants, and then broke away from all control. As he was a sacred being, he might not be destroyed nor even injured. By direction of the high priest a fence of consecrated bamboos was hastily set up round him, but he made short work of the bamboos, and the high priest himself had a narrow escape of his life.

"His Sublime Grandeur" then fortunately made his way into a court of his palace, where he could be barred in. Just as a cat does during a fit, the elephant dashed himself furiously against the walls, trying to batter them down with his tusks, and at last inflicted such injuries on himself that he fell dead.

Now it would have been much kinder to His Sub-lime Grandeur if his attendants could have placed him under control during the period of his madness. It would not have lasted for any length of time, and the animal might now have been enjoying the luxuries of his royal home, and the king and court of Siam would not be wearing the garb of woe.

That semi-worship should be offered to an elephant in Siam, may seem absurd enough to us in England. But really, when we recall the history of the great African elephant "Jumbo," I do not think that we can fairly laugh at the Siamese. The strangest part of the Jumbo-worship is, that it sprang up like a mushroom, in a single day.

There were four elephants in the Zoological Gardens, two from Africa, and two from India, the latter having been brought by the Prince of Wales after his tour in India, in 1875-6. Of the two African specimens, the male, named "Jumbo," was obtained by exchange from Paris, and the female, "Alice," was purchased in 1865.

Of all these creatures, the Indian specimens are the most generally interesting, being playful, and so gentle that they are quite pleased when the keeper's children enter their enclosure. Now, Jumbo, though a good-tempered and docile beast enough, had for some time been so uncertain in his temper, that only his keeper, Scott, dared to enter the cage alone.

Temporary madness does not exclusively belong to the male elephant, as is generally supposed. With him, it is almost sure to take place after he has attained adult age.

The Indian magnates are so well aware of this fact that in order to gratify their love of a peculiar department of sport, akin to the bull-fights of Spain, and the badger-drawing, bear-fighting, and dog and cock-fighting, which until lately disgraced our own country, they keep a number of adult male elephants for the purpose of fighting.

Elephants are mild enough except when in the state of "must," as this peculiar condition is called, and when two "must" elephants are placed in proximity to each other, how they fight is admirably told by Dr. W. Knighton in his "Private Life of an Eastern King."

Mr. Davis, the American agent who came to buy Jumbo, mentioned to me that out of the great number of elephants which had been possessed by the firm for which he is acting, some of the most dangerous were females. Few of my readers may be old enough to recollect "Madame Jack," the elephant which took an important part in several plays at the Adelphi Theatre, many years ago. She, like others, went mad, killed her keeper, and, I believe, several men besides, and then had to be destroyed.

Mr. Davis told me that the first sign of the distemper is that the elephants begin to play with something that takes their fancy, and become so excited that they do not obey their keepers. So that for all elephants, male and female alike, the means of restraint ought always to be at hand. We will now return to Jumbo's life in this country.

To naturalists he was of more importance than either of the others, as he was the first example of an African elephant ever known to be imported into England. To myself in particular he was a singularly interesting creature, and I have watched him at intervals since he was no larger than a Shetland pony.

Indeed, so anxious were his owners and keepers, that Professor W. H. Flower, the President of the Zoological Society, stated that he would have consented to Jumbo's removal even if nothing had been paid for him. More than this, Mr. Bartlett, who has had a life-long experience of elephants, was obliged, many months ago, to apply for means of instantly destroying the animal if he should break out into madness.

As Dr. Sclater, the Secretary, very forcibly remarks of the great establishment possessed by Messrs. Barnum, Bailey, and Hutchinson: "In so large an establishment, any animal under temporary excitement can be withdrawn from exhibition and placed in seclusion, which there are no adequate means of doing in the Zoological Gardens."

Several correspondents stigmatized this "excuse," as they were pleased to call it, as feeble and irrelevant. To my mind, it is simply convincing. Surely it must be kinder to Jumbo to place him among friends who

can restrain him during the short and distinct
intervals of excitement, and so enable him to enjoy
a long life of petted luxury, than to destroy him in the
first heyday of youth.

There is a stock argument very much in use by the
advocates of total abstinence, to the effect that man
is the only animal that will drink intoxicating liquors.

It is a pity that they should employ such an argument, or rather, illustration, for most animals will
indulge in stimulants when they can obtain them. In
one of Charles Dickens's letters there is an amusing
account of a Newfoundland dog that used to go to a
public-house every morning, and have his pint of beer
" drawed reglar, as if he was a brickmaker."

Wearied horses can be rendered capable of continuing their journey by the administration of a quart
of ale. Cows have more than once been known to
drink home-brewed ale that had been set outside the
farmhouse to cool, and to play the most ludicrous
antics in consequence of the indulgence.

The elephant is no exception to the rule, but is a
most determined toper whenever he can find an
opportunity.

A well-known writer and lecturer on total abstinence
lately cited "Jumbo" as a proof that the largest and
strongest quadruped in England was a teetotaler.
Had he made himself acquainted with the habits
of the animal, he would have found that " Jumbo,"
like all of his kind, is inordinately fond of any
alcoholic liquid, preferring whiskey or any other
liquor.

The Indian mahouts, when they have to teach
their animals any new accomplishment, always reward
them with arrack when they succeed, and the promise
of a bottle of arrack will always induce an elephant
to exert itself to the utmost

The mode in which the animal drinks a bottle of beer, wine, or spirits is very curious.

The cork is half drawn, and the bottle handed to the elephant. The animal puts the bottle on the ground against a wall or tree trunk, holds it firmly with one of its fore feet, grasps the cork with the finger-like appendage at the end of the proboscis, and twists it out in a moment.

Then it takes the bottle by the mouth, and gradually tilts it up until all the contents have been transferred to the trunk. Then it gives the empty bottle to the keeper, puts the end of its trunk into its mouth, blows the whiskey down its throat, and holds out its trunk for another supply.

A rather ludicrous example of the fondness of the elephant for spirits was lately exhibited by two of Mr. Barnum's elephants.

They had taken cold, and had a fit of the shivers. A gallon or so of whiskey administered to each of them speedily set them right. Next morning they were quite well, but as soon as their keeper came in sight they began to shiver violently, in hopes of obtaining another dose of whiskey.

Some years ago, an elephant, which belonged to a travelling company, was housed for a night in the stable of a hotel. Next morning the elephant was gone, and no one had heard or seen anything of him. That he should have been stolen was not likely, for the thief could make no use of him, and how so huge a beast could have concealed himself was a mystery. The country was scoured in vain all day ; but in the evening, a servant, who had occasion to go to the wine-cellar of the hotel, there found the elephant very quietly reposing among the bottles. The animal had evidently been attracted by the scent of the wine, and with the soft, noiseless tread of its kind, had found its way up the great stairs of the hotel, through the hall, and so into the wine-cellar.

Much of the interest excited by Jumbo is due to his enormous size.

We read in many books of travel that elephants are found from twelve to sixteen feet in height, while some writers have even ventured upon twenty feet.

Now an elephant of that height would be so enormous that the tallest giraffe ever known would need almost an additional yard of height in order to look over the elephant's shoulder.

In India, the elephants used for riding are on an average about eight feet in height, and may be compared to men of five feet six inches. Many are little more than six feet high, while an elephant of nine feet is considered a large one, and a nine or ten feet animal is about equivalent to a man of six feet two inches. Any animal that passes ten feet takes rank among giants. Every inch added to the height causes a proportionate increase of bulk, so that when Jumbo stands by the side of one of the Indian elephants, he looks like a dray horse compared to a Shetland pony.

Figures alone give but a poor idea of bulk.

In order to realize the gigantic dimensions of Jumbo, measure eleven feet in height on the side of a room, and then measure fourteen feet lengthwise; then picture to yourself an elephant of that height and length, and you will form some idea of the proportions of Jumbo. Perhaps even those proportions may be exceeded in time. He is yet but a lad, according to the duration of elephant life, and if he were to attain another six or seven inches in height, and gain another ton in weight, I should not be surprised.

For years Jumbo was an inmate of the Zoological

Gardens, growing rapidly, and in course of time helping the keeper to amuse the younger visitors by carrying them on his back. Even their elders did not disdain a ride on so vast an animal, as I can vouch from personal experience. Still, except for his size, Jumbo created no particular interest, and the public cared no more for him than for the other elephants.

Then a report was bruited abroad that an American agent was negotiating for the purchase of Jumbo, and the public naturally thought that the Zoological Society would do itself harm by parting with the largest elephant that had ever been known in Europe.

Still, beyond a few remonstrances, no great objections were made to Jumbo's removal. Indeed, the generality of the visitors to the Zoological Gardens did not even know the animal's name, nor, indeed, could most of them distinguish one elephant from another. Even on the eve of his departure from England I heard several persons assert that the female Asiatic elephant (" Suffa-Kulli ") was Jumbo, while, on the other hand, there were quite as many who pointed out Jumbo as one of the Indian elephants.

One morning, however, there appeared in one of the daily newspapers a vivid and dramatic account of an attempt to take Jumbo out of the Gardens, so as to accustom him to the road.

We were told how he suspected a trap, and bewailed his hard lot; how he knelt to his keeper, caressed him, and in all but human words besought for restoration to the home of his childhood. We learned how the other elephants from within their houses responded to his piteous appeal, and how they all rejoiced together when he returned among them.

M

As by magic, a Jumbo literature sprang up. The president, secretary, superintendent, and other officers of the Society were inundated with letters, mostly composed of vituperative epithets. Even persons like myself, who have no connection with the Society, but were known to take an interest in animals, received letters from all quarters on the same subject. None of the writers seemed even to conceive the idea that the officers of the Society were likely to understand their own business, and would not part from such an animal without very good reasons for doing so.

Then the Jumbo-worship set in. A Jumbo Rescue Fund was started. Presents of the most fatuous description were showered on the animal. Visiting cards with "farewell" were attached to his box, which was simply covered with farewell messages in pencil.

That basket after basket full of hot-house grapes should be given to him we can understand, though the grapes would have been better employed if given to sick poor who needed them, and who would not have eaten the baskets as well as the grapes.

But it is scarcely possible to conceive how many human beings could have been so ignorantly foolish as to present an elephant with several boxes of cigars, packets of snuff, a leg of mutton, and six dozen oysters.

From their nature these gifts seem to have been presented by donors of the male sex. But feminine presents are even more absurd than the masculine. No one, however imaginative, would have thought that a widow's suit should be sent to Alice, to be worn on Jumbo's departure. Or that numbers of ladies would send their photographs for Jumbo's consolation during his absence from them. Or that measurements of his travelling box should be taken, so that it might be decorated with wreaths of flowers.

Can anything be much more absurd than such conduct as the following? "On Wednesday, we saw a lady weeping copiously in the Gardens. With streaming eyes and a moist handkerchief, she was testifying to the violence of her grief, inveighing against the brutality of allowing Jumbo to *catch cold in his legs!*"

It is impossible not to recall Trinculo's soliloquy on discovering Caliban: "Were I in England now, as once I was, and had this fish painted, not a holiday fool but would give a piece of silver. When they will not give a doit to relieve a lame beggar, they will lay out ten to see a dead Indian."

One case, however, outdoes in absurdity all the previous instances of human folly. Here is an extract from the first column of a daily newspaper, the names being suppressed:—

"On the 27th ult." (*i.e.,* February, 1882) "at ——, the wife of H —— B——. Esq., of a son and heir (Jumbo)."

Here each sex is equally responsible, as both husband and wife must have concurred in saddling their unfortunate "son and heir" with a name that will afflict him during the whole of his life, and, if he should go to a public school, will be a perpetual torment to him. "Tristram Shandy," of Sterne, or Lord Lytton's "Anachronism," "Pisistratus Caxton," were nothing in comparison of "Jumbo B——."

Then legendary history was foisted upon Jumbo, and among other fables we were told that "he had been on exhibition at the London 'Zoo' for nearly sixty years, and that upon his back Queen Victoria and the royal family and thousands of children have ridden."

Now, all this outburst of Jumbo-worship was the

work of a few days, the rest of the animal's life in the Zoological Gardens not exciting the least enthusiasm in the public mind, even among those who had ridden on his back when boys and girls, and had in after years lifted their own children into the familiar howdah.

Yet to all naturalists the years in which he passed from infancy to adult age were full of interest, and to none more so than to myself.

Some twenty-five years ago I stated ("Illustrated Nat. Hist.," i., p. 739) that I believed the African elephant to be quite as well fitted for the service of man, and that the reason why it was not captured and tamed might be found in the inferiority of the negro race when compared to the Aryan. Many of the elephants which were employed in the days of ancient history were undoubtedly of the African species, as were those highly-accomplished animals which are stated to have walked along a set of ropes, carrying a companion in a litter.

That an African elephant should be brought to England was an epoch in Zoology, especially as the animal was very young, and might therefore be expected to live sufficiently long to enable its disposition to be carefully studied. He was then scarcely as large as an ordinary Shetland pony, and, up to the present time, when he is eleven feet in height at the shoulder, and weighs some seven tons, he has proved quite as gentle and docile as any of the Indian animals. His compatriot "Alice" has also proved herself as intelligent and capable of subjection to man as either of the two Indian elephants.

Yet the art of elephant taming has not been practised in Africa for many centuries. The natives can kill them by catching them in pitfalls, or by the "drop-trap," *i.e.*, a device by which a log of wood, armed with a poisoned spike, or a long, double-edged blade, is dropped upon them from a height.

Some tribes, more courageous than the rest, can hunt down the animal, and mob it to death, flinging spears at it until the creature dies from weariness and its multitudinous wounds.

Bravest of all are the Aggageers, so well described by Sir S. Baker.

They hunt the elephant in pairs, one being armed with a long, straight sword, the edge of which is kept as keen as that of a razor, and the other being unarmed. When they hunt, both mount the same horse, the armed man being behind.

Picking out an elephant with good tusks, they ride towards him and attract his attention. The man with the sword then slips off and hides himself under any convenient bush which they may pass. His companion then irritates the elephant, until it charges him. The horse, which is always of the swiftest kind, and carefully trained for the purpose, intentionally keeps just so far in front of the elephant that the latter thinks of nothing but catching it.

In course of the chase, the horseman passes close by his comrade's hiding-place, the elephant being too much excited to detect him. As the great beast passes, the hunter steps from his ambush, and with a single blow severs the tendon of the heel, which in the elephant is close to the ground. The animal is instantly rendered powerless, and can be killed without the possibility of resistance.

Even if the tendon be only partially severed, the next step is sure to snap it.

Yet in spite of the ingenuity of inventing such a feat, and the cool daring by which it is accomplished, no Agageer ever dreamed of taking the elephant alive. Nor would the Zulus, bravest of the brave as they may be, and utterly reckless of their own lives, attempt such a feat. They have been known to catch a lion alive, at the command of their king, but the very idea of taking an adult elephant alive would not have entered the head of Chaka himself.

Now, both the Agageers and the Zulus are of a much higher type than the negro, and it is therefore not at all wonderful that the negro cannot tame the elephant when tribes which are far superior to him fail to do so.

In general formation, both species very nearly resemble each other, and if the skull were removed, it would not be easy to decide whether the rest of the skeleton belonged to the African or Asiatic species. The form of the head is, however, very different, especially in the living animal.

In the first place, the enormous comparative size of the ears in the African species renders it so conspicuous that even when the animal is at rest and the ears are pressed closely against the head, there is no possibility of mistaking one species for the other. These ears, however, are best seen from the front, when the elephant is excited. In such a case, they stand out boldly on each side, looking like a pair of huge black wings.

Looking at the two species in profile, it is easy to see that the forehead of the African is convex, while that of the Asiatic is concave. Looking at them from the front, the head of the African narrows below the eyes, and then widens again, very much like that of the hippopotamus.

Its form is due to the manner in which the tusks are set in their sockets.

In the Indian species, the sockets run nearly parallel to each other, so that the skull is of tolerably equal width.

Both species have the peculiarity that if, when

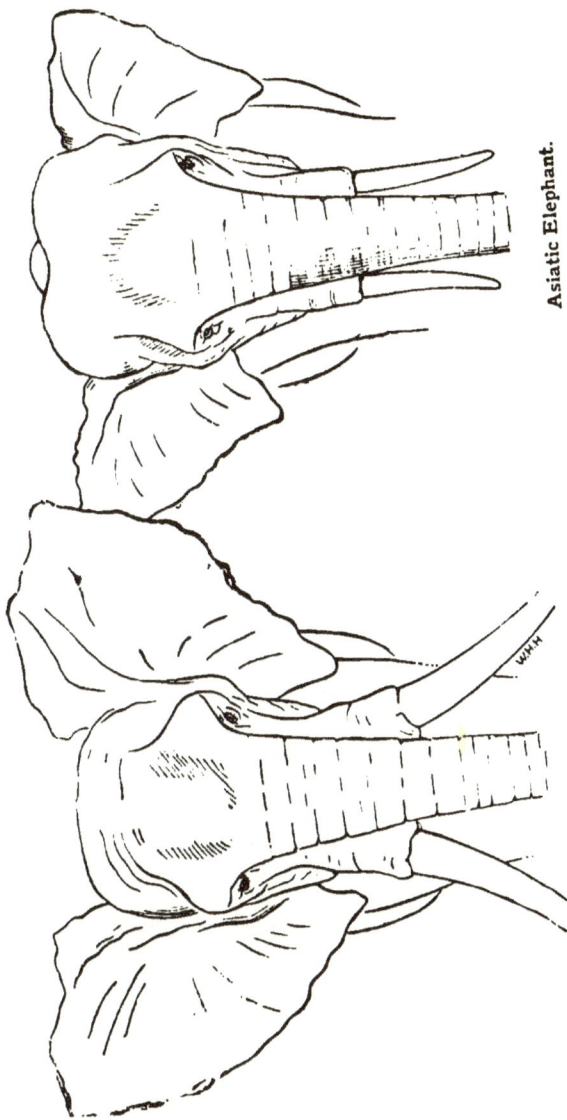

Asiatic Elephant.

African Elephant.

wounded, they once fall, they never rise again. The lions, tigers, bears, and even the buffaloes, will spring to their feet even when mortally wounded, and often kill their slayer with their last struggles. But the dying elephant "subsides like a great hayrick," to use Mr. Sullivan's words, and expires so gently, that the hunter is often uncertain whether the animal be dead or merely resting.

Ivory workers often find bullets imbedded in the tusks. They have struck the root of the tusk, which is hollow, and filled with pulp, and have been gradually carried towards the tip by the growth of the tusk. In the ivory turners' department of the Crystal Palace, there are some very curious examples of imbedded balls. In one case, the track of the ball is marked by a tunnel of bone extending across the base of the tusk, the ball itself having passed towards the point.

As is the case with the whale tribe, the brain of the elephant is very small in proportion to the size of the head, and is so deeply sunk in its bony outworks, that the hunter must aim as accurately as if shooting at a sparrow.

It is quite common to find in the skulls of slain elephants the marks where bullets have passed through the honeycomb-like mass of bone which surrounds the brain, and where the damage has been repaired by Nature. There are several such specimens in the College of Surgeons.

If the African elephant cannot be induced to turn his side towards the hunter, the only hope of the latter is to aim at one of the legs so as to disable it.

The late Gordon Cumming once owed his life to an accidental shot which broke the elephant's leg and rendered it powerless. As it could not stir, he, knowing very little of anatomy, tried to find its

vulnerable points by shots at short distances. After
receiving with comparative indifference ball after

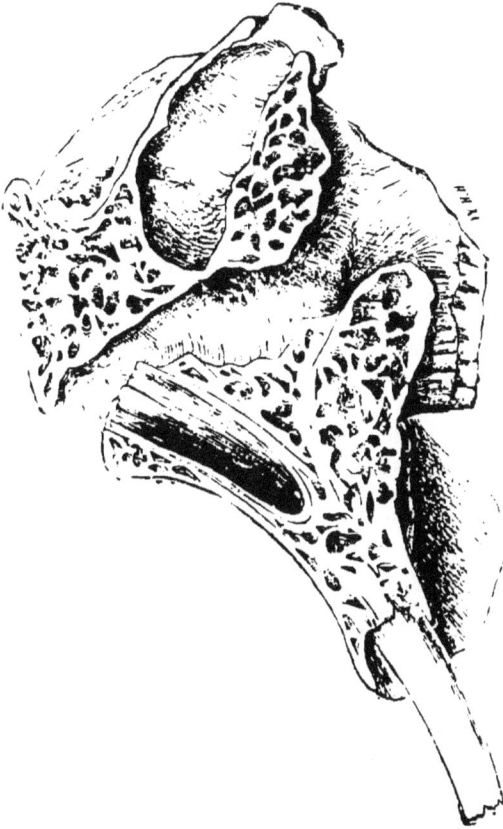

Section of Skull of Elephant, showing the Small Brain Cavity, and Honeycomb-like Nature of Bone.

ball, it sank dead from a shot which took effect
between the eye and ear.

He incurred much blame for the heartless cruelty
of this proceeding, but in reality he saved pain to

hundreds of other elephants, not to mention human lives.

It was far more humane to learn how to kill an elephant instantaneously with a single shot, than to allow the animals to be caught in pitfalls and transfixed by a stake, or to be slowly tortured to death by countless spear wounds.

Also, with the flesh of the elephants he fed whole tribes of starving natives, and so he brought the ivory into the market with the least possible pain to the animals, and the greatest possible good to the various native tribes, which depend largely on the elephant for their subsistence.

Both species have a similar gait. When they walk they do not, like the horse, move the feet alternately and diagonally, but walk alternately with the feet of each side.

Moreover, instead of bending the leg at the so-called "knee" and "hock," as the horse does, and furthermore bending it again at the pasterns, the elephant scarcely bends its legs at all, but swings them forwards and backwards, planting the heel first on the ground, just as man does. In fact, the elephant is, in our modern athletic slang, "a fair heel and toe pedestrian."

A glance at the skeleton will show the reason for this gait. In the elephant, the "cannon bones," or "shank bones," *i.e.*, the middle metacarpal bone of the fore foot and the middle metatarsal bone of the hind foot are not lengthened as in the horse, and the entire foot is brought close to the ground, all five toes resting on it.

This peculiar structure of the legs enables the elephant to use them as offensive weapons. It does not kick with its hind legs like the horse, nor strike, boxer fashion, with its fore-feet, like the stag, but it

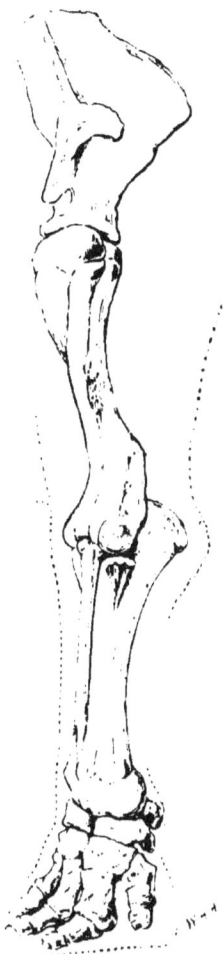

Bones of Fore-leg   Bones of Hind-leg.

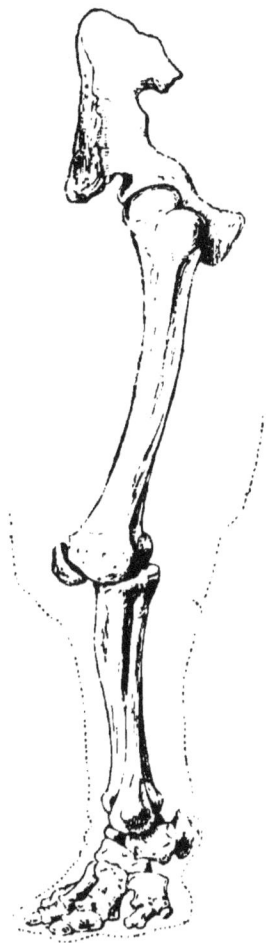

hustles its foe backwards and forwards under its body, kicking it forward with the hind feet, and then backward with the fore feet. In this way it has been known to destroy a wild boar and a tiger, and in both cases the elephant was a female which was defending her offspring.

It is strange to see how artists ignore this structure, even when they are engaged in scientific work. I have now before my eyes a well-known zoological diagram for schools, in which the elephant has knee, hock, and pasterns just like a horse, and to make matters worse, is standing with the pastern of one hind foot gracefully bent!

I may here mention that the footprint of an elephant designates the size and serves to identify the animal.

It is found by measurement that twice the circumference of the foot is equal to its height at the shoulder. Now, the circumference of "Jumbo's" foot slightly exceeds five feet six inches, so that his height is a little over eleven feet.

The identity of the animal is shown by the lines which cross and recross each other in the sole of the elephant's foot, just as do the lines of the palms of our hands, and which are imprinted on soft ground. When hunters track an elephant, they copy these lines, and so are able to adhere to the "spoor" of the same animal, even when it has been mixed with the footsteps of many others.

These great feet, which can crush a tiger into a jelly, and which have to support a weight which is measured by tons, are as silent in their tread as those of a cat. All elephant hunters know that elephants can glide noiselessly through thick forests, where even the barefooted savage can scarcely tread without betraying his whereabouts.

Even upon the highroads of this country, where the shod hoof of the horse is audible far off, the elephant swings his mighty bulk along without apparent effort, and so silently that "the blind mole may not hear a footfall."

Huge as it may be, no creature is so difficult of detection. Dr. W. Knighton, the Cingalese elephant hunter, tells me that in the forests of Ceylon you may be standing within a couple of yards of a nine or ten feet elephant, and not be able to distinguish the animal from surrounding objects. Its legs are just like tree trunks, and its brown body merges so imperceptibly into the sombre forest shadows, that the eye is incapable of discerning it.

Both species are playful, and are even fond of toys. In one case, a large wooden ball was given to the elephants. But they became so excited with their toy, hurling it about as if it were shot from a cannon, that the keepers were obliged to remove it.

Both species practise a most curious mode of avenging themselves when angered.

In Mr. Baldwin's work on African hunting, it is mentioned that fully half a mile from any water a tolerably large crocodile was found, hanging in the fork of a tree about ten feet from the ground. The natives seemed to be familiar with this strange position for a crocodile, and said that the reptile had been put there by an elephant.

They stated that when the elephants wade into the lake (Nyami) for bathing purposes, the crocodiles are apt to worry them and bite their legs. Sometimes, when an elephant is annoyed beyond all patience, it picks up the crocodile in its trunk, puts it among the branches of a tree, and leaves it there.

The truth of this curious story is corroborated by

the behaviour of an Indian elephant, very inappropriately named Pangul, or Fool.

The animal knew perfectly well the weight of the burden which he had to carry, and if he were overloaded, either refused to stir, or shook off his load by wriggling his skin.

One day an officer was trying to overload him, and became so angry at seeing the load repeatedly thrown off, that he flung a tent-peg at the elephant. Pangul took no notice at the time, but a few days afterwards he met his persecutor alone. Pangul immediately picked him up with his trunk, put him among the branches of a large tamarind tree, and left him there to get down as he could.

Gigantic as the elephant may be, it is horribly afraid of any small quadruped. A kitten which happened to stray among some elephants drove them half mad with terror, occasioning as much unreasonable consternation as a cockroach or mouse in a drawing-room full of ladies.

Yet, an elephant has been known to take a fancy to a cat, or rather, the cat took a fancy to the elephant. She had a fixed idea that his back was a sleeping-place expressly designed by nature for her, and on his back she *would* go. At first, the elephant took her off his back and put her out of his cage, but as fast as he put her down in front, she slipped round and climbed up his hind quarters again. So the elephant let her have her own way, and soon became quite attached to the cat.

The reader may remember that I have alluded to the assistant male elephants which play the part of the Philistines to the captive Samson, just as the koomkies take the part of Delilah.

These assistant elephants are as carefully trained to fight as our modern boxers, and, as with man,

size and brute strength are of small avail before
practised skill. The professional fighting elephant
knows beforehand every move in the game,—when to
bump his antagonist against a tree and thrash him
on the neck with his trunk before he can recover
from the shock, when and how to use his tusks, and
when to charge with his whole weight against his
adversary.

Some of these trained elephants have been sent to
Africa for the double purpose of capturing African
elephants and of showing the more intelligent tribes
how to take elephants alive instead of merely killing
them for the sake of the ivory.

The account of their transit is a very interesting
one, but too long to be given in full. There was
much the same difficulty in embarking them as was
found with "Jumbo," but there was much more
difficulty in landing them.

Owing to the peculiar shore of Zanzibar, the ship
could not approach within two miles of land. At
last it was decided to lower one of them, poetically
named "Budding Lily," into the water, and induce
her to swim ashore.

So she was slung over the side, and let gently into
the sea, with the mahout on her neck. Now elephants, when bathing, are rather fond of playing a
practical joke on the mahout. They sink themselves beneath the water so as to give the mahout a
sound ducking, while they can breathe through the
end of the proboscis, which is held out of the water.

"Budding Lily" played this same joke, but when
she rose to the surface became alarmed and tried to
scramble on board again. The captain of the ship
sent a boat, which tried to tow the elephant land
wards, but the animal was too strong, and dragged
the boat back to the ship's side, up which it vainly
attempted to climb.

After more than an hour had been thus wasted the elephant suddenly comprehended the situation, and swam towards shore, accompanied by the boat. Some sand-banks on the way afforded it resting places, and in about four hours after leaving the ship the first Asiatic elephant set foot on African shore.

There was little trouble with the other elephants, for they took courage from the conduct of their companion, and swam ashore after her.

One of the elephants of the Jardin des Plantes, of Paris, used to play an absurd trick with the visitors. She would sink herself until only the tip of the trunk projected from the water, and she was thereby rendered practically invisible. Then she would send a torrent of water over the spectators, who could not imagine where the deluge came from.

That elephants should be such admirable swimmers seems very remarkable, and especially that they should have such propulsive power as to drag back a boat fully manned. The capability of sinking or rising at will in the water is equally remarkable, and is owing to the power of contracting its body so as to render it heavier than an equal bulk of water.

Activity, again, seems no characteristic of the elephant. Its apparently stiff and ungainly legs, which can swing some fourteen or fifteen feet at each step, although well enough adapted to carry the huge body along at a swift pace, appear to be totally inadequate to perform feats of activity.

Yet the elephant can climb rocks where one might think no animal but the goat would venture. It can slide down a steep hill just as a "coaster" slides down a snow-clad declivity on his sledge, and can guide or check its progress with equal skill. I have seen an elephant stand on its hind feet, or fore feet, or on the feet of one side, or on the fore foot of one side and the hind foot of the other, and all the time mounted on a wooden cylinder not large enough to support all its four feet when placed together.

Yet, though it can swim so well, and can so easily ascend and descend precipitous slopes, the elephant is utterly powerless in mud of any depth. Should it by chance stray into a quagmire it becomes frantic with terror, utters screams of mingled fear and anger, its eyes start from its head with fright, and its proboscis feels in all directions for something firm on which it may stand.

When an elephant is in this predicament, the mahout slips off over the animal's tail, and runs away as fast as he can. Did he not escape in this way, the elephant would be sure to pluck him from its neck, and place his body under his feet so as to form a support for its weight.

The only plan by which an elephant can be relieved from this awkward position is to approach as near as is consistent with safety, and to throw logs, planks, or branches of trees within reach of the proboscis. The elephant immediately seizes them, and places them one by one under its feet until it can stand firmly, and by continuing the process, makes a road by which it may regain dry ground. As soon as it has done so, the mahout resumes his place on the animal's neck, and can safely guide it as before.

Cautious as is the elephant in trusting its vast weight to anything which may seem to be unable to support it, the animal's wonderful power of balance enables it to step for almost incredible distances from one foothold to another.

First, it surveys the intervening distance, and carefully takes the measure of it. Then it leans forward, and stretches its proboscis forwards so as to test the strength of the next foothold. Guided by the

N

proboscis one fore foot is then pushed forward until it obtains a hold.

Next follows the hind foot of the same side, and then the fore and hind feet of the opposite side are gradually transferred to their new situation, the ever restless proboscis always acting as pioneer of each step.

No one who has not seen it can realize the marvellous delicacy of the whole proceeding, or the perfection of balance shown by the apparently ungainly animal.

Judging by appearances, the elephant is about the last animal in the world which we should have thought to be swift of progress on land and in the water, a rock-climber, silent of tread as a cat, almost invisible among trees, and capable of slipping through dense forests without shaking the boughs or producing a sound that would betray its presence. Still less could we expect that it should be able to perform the extraordinary feats of agility which have already been mentioned.

Yet it does all these things, and, most wonderful of all, its vast strength, its powers of intellect, and its other great gifts are made to be subservient to man.

I was never more impressed with the truth of the passage quoted at the beginning of this article, than by watching the mode in which the enormous animal was rendered helpless by man, into whose hand all living creatures are delivered.

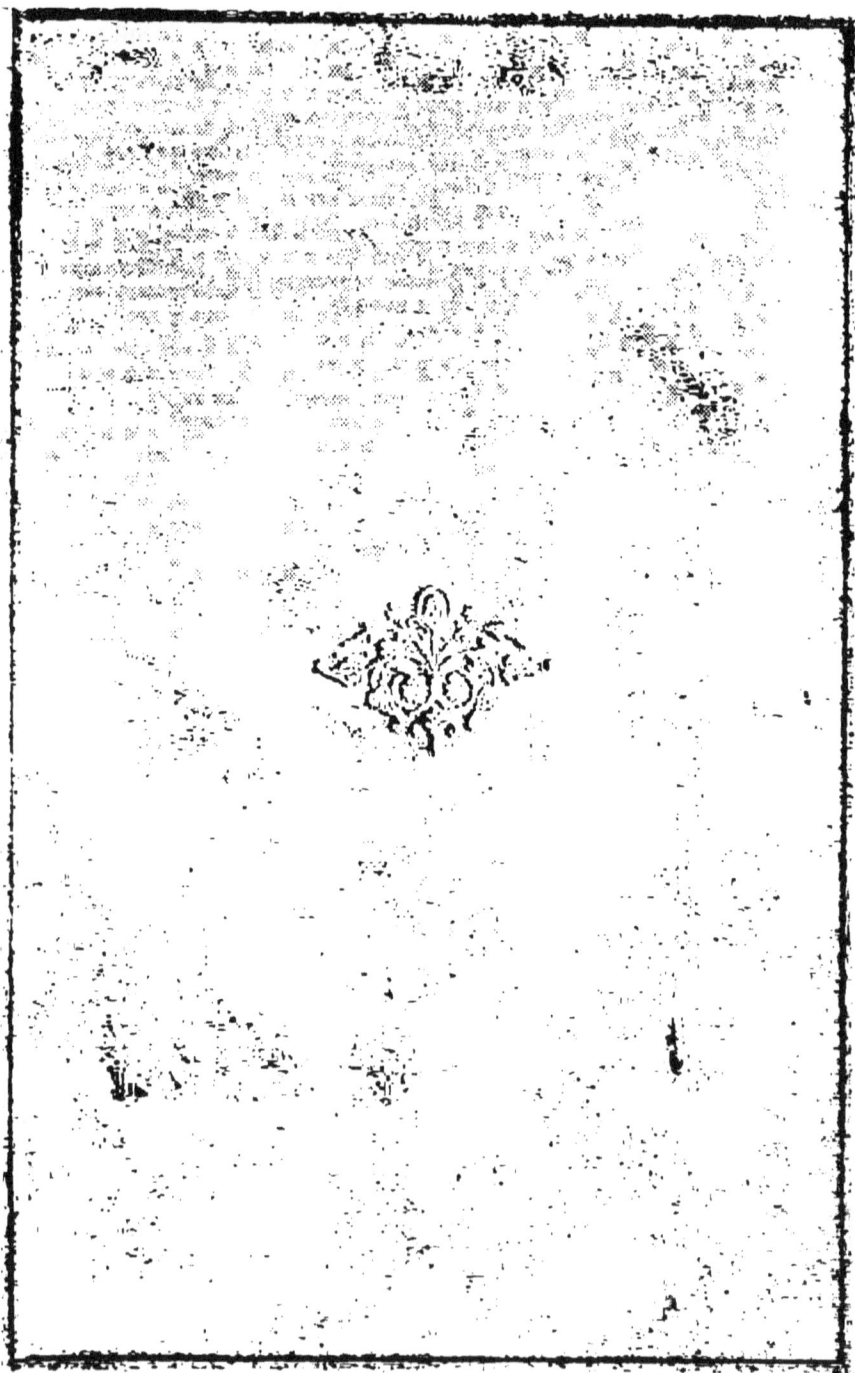

www.ingramcontent.com/pod-product-compliance
Lightning Source LLC
Chambersburg PA
CBHW021709210326
41599CB00013B/1578